Chicago Public Library
C
P
L
REFERENCE
Form 178 rev. 11-00

Writing and Designing Manuals

THIRD EDITION

Operator Manuals

Service and Maintenance Manuals

Manuals for International Markets

Writing and Designing Manuals

THIRD EDITION

Operator Manuals

Service and Maintenance Manuals

Manuals for International Markets

Patricia A. Robinson

Ryn Etter

CRC Press

Boca Raton London New York Washington, D.C.

Library of Congress Cataloging-in-Publication Data

Robinson, Patricia A., 1948–

Writing and designing manuals : operator manuals, service and maintenance manuals, manuals for international markets / Patricia A. Robinson, Ryn Etter.—3rd ed.

p. cm.

Includes bibliographical references and index.

ISBN 1-56670-378-6 (alk. paper)

1. Technical writing. I. Etter, Ryn. II. Title.

T11.R635 2000

808′.0666—dc21 99-08627

CIP

International Standard Book Number 1-56670-378-6
Library of Congress Card Number 99-086267
Printed in the United States of America 1 2 3 4 5 6 7 8 9 0
Printed on acid-free paper

Preface to the Third Edition

Like its predecessors, this edition of *Writing and Designing Manuals* is written for the technical writers, editors, and graphic artists who labor in the trenches doing the best they can with too few resources and too-tight deadlines. Over the 15 years since the first edition appeared, much in the profession has changed, but the central problem remains: how to get needed information to users in a form they will actually *use*. As in previous editions, we have relied on the experience of hundreds of technical writers, editors, and graphic artists from a broad spectrum of industries. Some came to seminars, some we met consulting, but all of them gave us the benefit of their perspectives. The suggestions in this book are based on what works in the real world, not some idealized academic construct. We asked the people doing the job what works for them, and this book represents their collective answer.

As in previous editions, we have organized the book to reflect the major steps in manual production: planning and understanding the context, analyzing the user, choosing organizational and writing strategies, designing a format, and developing graphics. Also included are chapters addressing special topics: safety warnings, service and maintenance manuals, manuals for international markets, and managing the technical publications function.

Reflecting the changes in the world since the last edition, we address the impact of new production technology, the rapid growth of the Internet as a way to do business, changes in product liability law, and the emergence of the European Community, with its implications for manual writers. We have included many new examples taken from manuals that reflect the increasing quality to be found in the profession. When we first started in this business, it was not uncommon for the product manual to be a couple of pages of typewritten text. Now, thanks to computer technology and the growth of trade journals such as *Technical Communication*, even tiny companies can produce slick, professional publications. We have grown along with the profession, and hope that this new edition of our book will continue to be a source of information and connection for technical communicators engaged in the messy, complex, frustrating, and fascinating business of producing manuals.

Patricia A. Robinson
Ryn Etter

The Authors

Patricia A. Robinson has been involved in technical writing for more than 20 years. After earning her Ph.D. in 1978 (in a different field), she joined the faculty of the University of Wisconsin College of Engineering where she taught technical writing until 1991. During those years she taught a variety of undergraduate courses, served as the first Director of the Technical Communication Certificate Program, and was named Manager of Campus Courses for her department. During her tenure at the University, the technical writing program grew from one report-writing course to a full certificate program. Dr. Robinson also served as program director and primary instructor for continuing education seminars for practicing technical writers and engineers, covering everything from effective speaking to managing the technical publications department.

Dr. Robinson is the author of *Fundamentals of Technical Writing* and coauthor of *Writing and Designing Manuals* and of *Effective Writing Strategies for Engineers and Scientists.* She has also published articles in various professional journals and presented papers at professional conferences. She is a regular speaker at product liability seminars on the role of instructions and warnings.

As owner of Coronado Consulting Services, LLC, Dr. Robinson consults widely for industry both in the U.S. and internationally, primarily in the area of instructions and warnings. She has testified as an expert witness in product liability cases. Past clients include Johnson Controls, GE Medical Systems, Alcoa, and many others.

Ryn Etter is a communications consultant and former full-time lecturer in the Department of Engineering Professional Development at the University of Wisconsin–Madison. She holds a Ph.D. and M.A. in English and a B.S. in journalism. She is a three-time recipient of the Outstanding Instructor Award from Polygon Engineering Council at the University of Wisconsin. Dr. Etter teaches technical writing, technical editing, and writing/designing user manuals, and is a frequent speaker at product liability and safety warnings seminars.

She served as faculty advisor in the university's Technical Communication Certificate Program and as advisor to *Wisconsin Engineer* magazine. She directed the College of Engineering first-year composition courses for all engineering students and served as a facilitator for the National Science Foundation Engineering Education Scholars Program. Dr. Etter has worked for over 25 years as a teacher, guest speaker, facilitator, writer, proofreader, book reviewer, editor, and consultant. She has also worked as a tech hand and light-board operator for a theater production company, a tractor driver, and a stonecutter for jewelry and sculpture.

Acknowledgments

Many individuals have contributed to this book by sharing their expertise and their materials or by allowing us to observe and work with their writers. We owe a special thank you to the following people: Cal R. Burnton of Wildman, Harrold, Allen & Dixon; John Conrads, formerly of Deere and Company; Dale Fierke, Tetra Pak; Charlie Freeman, Hewlett-Packard; John Gormley, Westinghouse Electric Corporation; David J. Howard, Clark Components International; Albert O. Hughes, FMC Corporation; Lila Laux; Bill Lichty, Scotsman; T. W. Loetzbeier, Mack Trucks; David S. Maslowski, Kartridg Pak Company; Jimmie Moeller, Gehl Company; Richard Moll, Department of Engineering Professional Development, University of Wisconsin–Madison; Geoffrey Peckham, Hazard Communication Systems, Inc.; Fred Rode, Outboard Marine Corporation; Pete Shelley, General Electric Medical Systems; Delmar Swann, E. I. DuPont De Nemours and Company; Stan Sweetack, Pierce Manufacturing Inc.; John Thauberger, Prairie Agricultural Machinery Institute; Ivan Thue, Prairie Implement Manufacturers' Association; George Winkleman, Delta International Machinery Group.

The materials for the book were collected in a number of ways. We have conducted in-house technical writing seminars for industries, visited service publications operations, lectured and taught in national and regional conferences and workshops devoted to product safety and technical writing, and served as private consultants. Individuals and organizations who have worked with us and/or whose materials provided the subjects and examples for this book include Acme Burgess, Inc.; American Honda Motor Company; American Optical Company; American Suzuki Motor Corporation; Atwood Mobile Products; Butler Manufacturing Company; Chrysler Motors Corporation; Clark Components International; Construction Industry Manufacturers' Association; CooperVision Surgical Systems; Deere and Company; Delta International Machinery Corporation; Devilbiss Company; Doboy Packaging Machinery; E. I. DuPont De Nemours and Company; Epson America, Inc.; Farm Implement and Equipment Institute; Fellowes Manufacturing Company; FMC Corporation; Ford Motor Company, Ford Tractor Division; Ford New Holland; Fulton Manufacturing Corporation; Gardner Denver; Gehl Company; General Electric Company and General Electric Medical Systems; General Motors Company; Gerber Products Company; Harley Davidson Motor Company; Hazelton-Raltech Incorporated; Hewlett-Packard; Honeywell Inc.; Huffy Corporation; Ingersoll-Rand Company; International Business Machines; International Harvester Company; J. I. Case Company; John Muir Publications; Joy Manufacturing Co.; Kartridg Pak Company; Kohler Corporation; Krones Inc.; Mack Trucks–Mack International; Madison-Kipp Corporation; Mantis; Martin Engineering; Norden Laboratories; Ohio Medical Products; Outboard Marine Corporation; Pierce Manufacturing Company; Prairie Agricultural Machinery Institute; Prairie Implement Manufacturers' Association (and affiliated companies of Canada); Rheem Manufacturing Company; Rosemount Inc.; Ryobi Outdoor Products, Inc.; Scotsman; Siemens Medical System; Silver-Reed America, Inc.; Teresa Sprecher; Rosemary Stachel; Sunstrand Aviation; TRW Ross Gear Division; Taylor Instrument Company; Technicare; Thern Inc.; Versatile

Corporation; Volkswagen of America; Wabco Construction and Mining Equipment; Westinghouse Electric Corporation; and Yahama Motor Company.

We wish to thank Beth Harper, government documents librarian at the University of Wisconsin Memorial Library, for her help in accessing and explaining European Union documents. She, the other Memorial Library reference librarians, and the staff at Wendt Engineering Library were invaluable in tracking down international documents.

Finally, we owe a special thank-you to our partners, Leslie Kramer and Marijane Ellington, for being patient and tolerant with us through the preparation of this manuscript.

for Gretchen Schoff

Contents

List of Figures

CHAPTER 1

Understanding the Context

OVERVIEW

Manual writers from many businesses have been our guides in creating this book. They keep asking questions that lead us to look for answers, and they have generously shared with us their favorite tricks of the trade. We have also benefited from academic research as technical communication evolves into an identifiable discipline and a distinct profession. Recent work in human factors and in computer interaction design, in particular, offers explanations and tools for creating manuals that are actually helpful to users.

From coffee-break anecdotes to classroom studies, one basic fact emerges. Whether writers are preparing manuals for copiers or paint sprayers, computers or trailer hitches, they encounter similar problems. And some of these problems occur before writers ever put pen to paper or stroke to screen.

This chapter discusses the context of manual creation — where and how the manual and the writer fit within companies today. Knowing where you are in a company's structure, how to get what you need from where you are, and how to plan before and as you write will help you create manuals people will really use — and not feel stupid while they're using them.

VALUE OF MANUALS TO A COMPANY

Operator and service manuals give verbal and visual instructions for the use and care of thousands of products, ranging from toasters and tractors to cameras and computers, not to mention the infamous VCR. Almost every product and piece of equipment except the simplest comes with such instructions.

These instructions are known as user manuals, operator (and service) manuals, owner manuals, or simply instructions. They are "how to" books for owners and operators of the products. Their unwritten goal, or hope, is to help the users not look stupid while using the manual or the product. (We will discuss how important this goal is in Chapter 2.)

Manuals have other, more obvious goals as well. They function for the company as instructions, product liability documents, and marketing and public relations tools all rolled into one.

Instructions

Most user manuals today contain the following sections:

- Assembly or installation
- Operation
- Maintenance (and storage)
- Troubleshooting
- Repair and service
- Parts

Very complex products, such as computer-controlled and electronics systems, heavy industrial machinery, or biomedical equipment often include a section of technical information, however simplified, and have separate manuals just for maintenance, service, and repair.

New products, for which no prototype manuals exist, and complex products, in which many subsystems interact, present especially difficult problems for writers. Such manuals demand that writers understand an astonishing array of mechanisms, processes, and procedures and be able to explain them with relative clarity and simplicity. Writers in these situations often find themselves revisiting physics, chemistry, and mechanics textbooks — all so they can understand the product and teach only the need-to-know information to the users. Writers, then, always ask themselves, "How can I best describe or show how this works?"

In creating clear instructions, writers are the users' advocates. Clear instructions benefit the company as well as the customer. Good manuals are likely to result in fewer customer service calls. Customers are less frustrated because they don't have to call, and the company is happier because fewer technical service staff are needed to answer questions. Manuals add value to the product and the company far beyond words and pictures on a page.

Product Liability Document

The consumer protection movement and the legal climate surrounding product liability law are of serious concern to manufacturers of any product. Manual writers need to look not only at the verbal and visual instructions that make up the manual, but also at the individual safety warnings placed throughout the document and on the product itself. This area of manual creation is so critical, we'll discuss safety warnings in manuals and on products in a separate chapter (Chapter 6).

One important area of product liability law we'll return to is the manufacturer's "duty to warn." In general, manuals must warn product users of hazards such as electrical fields, sharp blades, moving parts, shattering glass, dangerous fumes, chemical interactions, toxic substances, and flammable and explosive materials. Again, Chapter 6 will discuss the legal requirements in greater detail.

A user manual and the accompanying warnings that go on the product itself frequently become key documents in product liability suits. If the manual is well designed and worded, it may help protect the manufacturer against charges of failure to give adequate warning. In other words, the manual serves as part of the company's paper trail to show its consistent, from-the-beginning concern for safety. A company concerned about safety should include the manual writer in the risk assessment/hazards analysis team.

Marketing and Public Relations Document

A user manual sends a message to the buyer of your product. You can make this message a good one with an attractive, clear, and helpful manual. Such a manual is truly user-friendly: it invites and reassures users that the manual, and by extension the product, was designed with them in mind and that they are competent to use both.

Don't underestimate a user's need to feel comfortable and competent. The world of too many options and too complex technologies tends to produce anxiety rather than ease in most consumers. You're not likely to look forward to exploring the wonders of your new computer if the manual tells you everything except how to turn the blamed thing on. If the instructions for your new camera are so confusing you can't even put in the film without returning to the store (where a technologically advanced 17-year-old clerk gives you a bored and withering look), you already feel angry and dumb before you've taken one picture of Fluffy.

First steps, like first impressions, often are important. A manual can quite quickly cast a glow or a pall on your product. A manual's halo effect, then, influences how users see your product — and your company. If users think, "That's not so hard. I can do that," while reading the manual, they're also likely to feel the same about the product.

When you also stop to think about when many people actually consult a manual — when something goes wrong and they can't wing it — you'll quickly understand that a good manual helps reduce stress. Those who play with the product first and use — or perhaps, lose — the manual later are already frustrated when they finally get around to reading your instructions. Bludgeoning them with unreadable prose, dislocated structure, and obscure visuals will not calm them. It might glaze them into unconsciousness, but that is not the effect desired by most writers. The manual-as-last-resort should come to these users as a surprisingly helpful gift, not as another punishment.

Whether through invitation or desperation, then, a good manual adds value to your product. A manual is, really, a product in itself, and it reflects on the company like any other product. A good manual helps create customer satisfaction, repeat purchases, and fewer technical service, support, or help-line calls. If you're trying to convince your company of the value added by good documentation, talk to customers and count service calls before and after release of a new manual.

Manuals also work well as internal support within a company. They can become teaching tools to educate sales staff about the product, staff who in turn can use the manuals as marketing devices to pitch the products. When developed in sync with the product, a manual can also serve as a blueprint of what is needed or as a roadmap to keep everyone on track.

Products might not be sold based entirely on the use and beauty of their manuals — writers can only dream — but they and their companies certainly benefit from good ones.

Value Added and Beyond

The value added by good manuals becomes increasingly clear as companies move more and more into global marketplaces. Good product documentation is a requirement for European Common Market compliance, for example, and for International Organization for Standardization, or ISO, certification. For ISO 9000 quality assurance, for example, all but two of the 20 major system elements explicitly mention documentation as vital, and the other two elements imply it. (We'll discuss writing for international markets, and the problems of translation, in Chapter 8.) For U.S. companies doing business abroad, the publication does "live" with the product, and the product can have a very short life, even dying presale, if the publication doesn't work with it.

Within the U.S., specific industries also recognize good documentation as important. For example, the mandatory Hazard Analysis and Critical Control Points (HACCP) implementation that affects the meat, poultry, and prepared foods processing industry requires documentation of its seven principles (Galosich 48). Other industries require careful documentation as well.

Clearly, there is a continuing need for good documents, including manuals. Some writers wonder if the coming years, with more and more information moving online and on the Internet, will see an end to hard-copy manuals. We don't think so. Many products are not, in fact, computer related, and documentation online has its own set of problems and limitations. Most people of the world, in fact, do not have access to a computer, yet they still use a number of tools, devices, and pieces of equipment.

However, technical writers are being asked to handle multimedia production — CD-ROM, online, Web pages, e-mail support, etc. — as well as original hard copy. As technologies grow more complex, technical writers also become human interfaces between the majority of users and their machines. Within the computer industry, some experts are pushing for computer interaction designers — people who design computers for users, rather than for programmers — to be involved in product development from the beginning of the process. Sounds surprisingly like what technical writers have been arguing for with manual-and-product design: development in sync.

The publication (or the interface) should begin life with the product. The consequences of not doing so are already with us: "creeping featurism" and "the tyranny of the blank screen" that needlessly complicate products and bewilder users (Norman 172, 178) and "software apartheid" that separates and isolates the haves from the have-nots in terms of technology skills (Cooper 11). (See Chapter 2 for more discussion of these already-present consequences.) Who better to help us move more pleasurably into the information age than technical writers?

ROLE OF THE TECHNICAL WRITER

Technical writers, whether in a publications department of one or of many, frequently find themselves serving three unofficial but important roles in a company.

Traffic Controller

Quite often, all information roads lead through the technical publications department. In many companies, this department is the only place where sales literature, service and user manuals, parts lists, training manuals, engineering specifications and computer-aided designs, or CAD, drawings, graphics, video, film, and computer information all come together, or at least pass through. On a good day, traffic on this information highway is heavy but smooth. On a bad day, this publication intersection has only fast traffic and no stoplights — and no nearby exit ramps for overrun writers.

Translator

Technical writers often find themselves working as translators for the technologically impaired or technologically injured. Bruised by products that are beyond their sophistication, poorly designed, or aggressively user-unfriendly, consumers look to manual writers for help.

Troubleshooter

This role might be one of a company's unrecognized assets when it comes to product development. If brought into the process early enough — say, from the beginning — a technical writers can provide a fresh, often nontechnical yet technically friendly, questioning perspective. They can help with some design choices and safety analysis. Much like the software interaction designers mentioned before, technical writers can help engineers, programmers, and other research-and-development people see the user before they envision the product.

PLACE OF THE TECHNICAL WRITER: LOCATION, LOCATION, LOCATION

This section could also be called, "It's 10 o'clock. Does anyone know where the technical writer is?" If you haven't thought a lot about this question, you're not alone. As Nina Wishbow points out in "Home Sweet Home: Where Do Technical Communication Departments Belong?" academic studies have not explored the issue. Even Karen Schriver's encyclopedic *Dynamics in Document Design* addresses the question mainly from the perspective of the profession at large. However, Schriver's wonderfully titled section, "Practitioners without a Profession: 'Nobody Loves Me But My Mother and She Could Be Jivin' Too,'" could easily describe the place of technical writers in most companies.

Why so homeless? Money and time. Some companies have been so busy fire-fighting immediate crises they've had little time to plan careful documentation. Other companies still think of technical publications as an after-the-fact imposition, a soft and fuzzy add-on that pays back little yet costs a lot. This second-class status of technical publications, then, happens especially where manual production has grown rapidly over the years, with haphazard, unplanned responses to crises and needs, but no real plan for publications-and-product fit. Much like user interface for software design, manuals have often wound up being turned out as a product afterthought.

But in writing, as in real estate, location does matter. Often with no established, as-equally-important-as-other-departments home, or even when they have one, technical writers commonly face "overwork; lack of recognition; lack of space, equipment, the latest tools; lack of access to developers and other sources of information; lowered status relative to one's peers in other groups; lack of job depth or richness; need to transfer to another group to do tasks such as multimedia, user interface design, and so on" (Wishbow 29). We don't know how much these problems affect individual writers, manuals, or companies. We do know from talking with many, many writers that they do exist.

Location brings certain advantages and constraints to the writer's job. Some writers are outside the company, hired pens or have-laptop-will-travel consultants brought in for particular projects. Most often, technical writers within a company can be found in one of the following places:

- A stand-alone technical publications department responsible for all company products
- Separate technical publications units for each product, often located at widely separated geographic locations or within separate divisions
- Free-floating, attached only by a budget line to whatever department needs documentation
- Technical publications subsumed entirely under another division

In talking with writers and managers of writers, we've found this last location to be most common. If technical publications tends to be housed under another department, that department most often is engineering, marketing, or services/support. In these situations, technical writers many find themselves answering to bosses who are not writers or, in some cases, impatient with the concerns of writers, ill-equipped to meet their needs, or downright uninterested in the whole process of good documentation.

Most often, engineers, marketing, or service people are willing to help the technical writer. They just don't know how. Technical writers need backup from the boss, but they also need to develop effective networking skills on their own to get what they need.

A TECHNICAL WRITER'S SURVIVAL SKILLS

What is clear from the outset is that neither the "technical" nor the "writer" part of the job title is as important as the unwritten "people skills." Technical writers depend on good relationships with people within many departments — engineering, marketing, manufacturing, technical support, maintenance, and office services.

These people skills are so very important because the technical writer is seldom the subject matter expert. Instead, the writer relies on others for information that provides the substance of the manual. Good technical writers are quick studies and self-taught learners. They are also adept and informal interviewers, able to coax vital, need-to-know information from preoccupied sources.

Location in large part, therefore, demands that technical writers possess certain key survival skills to work effectively:

- Willingness to ask questions (and risk looking stupid)
- Interest in learning details of a project (even if some of those details never appear in the final manual)
- Ability to learn independently
- Sense of humor
- Flexibility and adaptability
- Capacity for careful listening
- Ability to write well under constant time pressures
- Recognition that perfection is not possible (and probably not desirable)

What the list boils down to is this: the ability to survive, even thrive, in chaos. Over the years we have talked with hundreds of technical writers, on their own and in companies large and small, in supportive and in hostile environments. We have not yet found a single one who thought her or his job was straightforward and easy. Instead, they describe it as a juggling act in which they are always short of two key elements: time and information.

THE WRITING PROCESS

The technical writer whose principal task is producing manuals has one of the toughest jobs around. It is tough because the writer is responsible for a crucial task — producing a manual that, aside from the product itself, is the company's main connection with the customer. Yet the writer seldom has the full authority needed to do that task well.

Indeed, many of the major decisions affecting the production of the manual, including both content (information) and schedule (time), are often made by persons in other areas of the company. Deadlines may be set by marketing to coincide with a new model period without regard to the complexity of the writing task. Information needed to meet those deadlines may be held up in engineering because of last-minute design changes. Yet the technical writer is expected to produce usable, accurate manuals, on time and within budget. And in the real world, writing manuals will never be accomplished the way you were taught to write papers in school.

Writing manuals the way you were taught to write school essays is both ineffectual and dangerous. You won't get a manual done, and you might lose your sanity.

The Writing Process (As Taught in School)

1. Make all basic document design decisions:
   - Content
   - Format
   - Schedule

2. Gather information
   - About product
   - About users
3. Prepare outline and list of visuals.
4. Write entire draft and prepare all visuals.
5. Edit and get approvals.
6. Print manual.
7. Distribute manual with product.

Instead, writing manuals will always look more like the following.

The Writing Process (As It Really Happens)

1. Receive assignment with nearly reasonable deadline.
2. Begin making some basic decisions.
3. Deadline moved up 2 weeks.
4. Try to get information from engineers. Receive spec sheet with illegible handwritten changes. Receive torn copy of competitor's brochure.
5. Try to get product. Receive outdated model with parts missing.
6. Deadline moved up 2 weeks.
7. Start to write anyway. Receive current prototype. Celebrate.
8. Overhear hall conversation about radical design changes in product. Scrap draft. Begin to read want ads.
9. Deadline moved up 2 weeks.

Writing as it is taught in school assumes two things: complete control and linearity of process — neither of which you have writing manuals for publication. (See References at the end of this chapter for books on creativity and the writing process.) What you do have, always and forever, is chaos and a deadline. Out of these you create a manual.

SO HOW MANY WRITERS DOES IT TAKE TO MAKE A MANUAL?

One or many: those are a company's choices. Whether the technical writing "department" is made up of one or many writers, each situation has its own advantages and disadvantages.

The Solo Writer

Small companies often assign one person to do the manuals for all products. If you are that solo writer, you will soon find that writing is only one part of the job. You may also have to do photography or scan it in, plan the art work, choose paper stock, edit, type, and desktop publish the entire manual. You will want to clarify, before you begin, how much control you have over the choices you'll have to make. Ideally, as a solo writer the control should be yours. If people you talk with seem vague about who decides what, try to claim the authority and the responsibility for the technical writing decisions. Convince others that you are the resident expert.

Advantages

As a solo writer, you have many opportunities to be creative. Because the majority of decisions will fall to you alone, you can approach the manual production job with your own vision of how the final manual will look, and you can make certain decisions without having to clear each step of the production with someone else.

We have met a number of solo writers who say that the autonomy they enjoy more than compensates for their many responsibilities. They also value the variety of tasks involved and enjoy the different kinds of people they work with. Most of all, they like having control over the project from start to finish.

Disadvantages

If you are a solo writer, your work will be the single bridge between the technical data about your product and the manual that reaches users. You will have to gather the information and create the schedule yourself. Manual writers who work solo often feel rushed, isolated, and pressured by their many responsibilities. They sometimes feel that other people, on whom they must rely for information, have little understanding of, or sympathy for, what it takes to put a manual together. In particular, they think other people often believe manual writing "happens" faster than it does, that it's a quick process. It's not, of course. But others may not understand this fact of manuals in particular, or of the writing life in general.

Making Solo Writing Easier

Much depends on your ability to handle solo writing in a professional manner, but you have the disadvantage of not having a writing team to lean on, either for moral support or advice. When you work alone, one of the most logical and understandable ways to get others to understand your work is to keep books on what you do.

For instance, you might begin by simply keeping a record of how many hours it usually takes you to create a manual page for a new product or how much time it takes to produce a computer-generated graphic. Even relatively uncomplicated bookkeeping can be a big help if you have some facts and figures to show that deadlines are unreasonable, costs are too high, or that one person cannot do the job alone.

Your needs as a solo writer are much the same as the needs of team writers. You need access to information and time to do the job. As you read the rest of the chapter on team writing and on information and time, you will find many suggestions that you can adapt to the solo-writing setting.

Consider, especially, ways you can perform the same functions a team leader performs in team writing. For example, you can do your own advance planning by:

- Arranging your own schedule of meetings with key personnel to collect information
- Asking for help from informal support teams or individuals (for work such as typing, drawing, planning safety messages, taking photographs)
- Developing a thorough outline
- Laying out steps in manual production
- Setting up a style and format guide or handbook so that your own writing procedures become standardized and easier to repeat from manual to manual

Team Writing

Manual writing is often done by a team of writers, especially if the product is complex. Such division of labor makes sense for a number of reasons: overall preparation time can be shortened, writers can develop special expertise with certain manual segments, and teams can include personnel from other company units (e.g., technical, research, product safety). The team-written manual also poses problems, particularly those of conceptual unity, team coordination, and uniformity of quality. Here are some of the pros and cons of the team-written manual and some suggestions for making team-writing efforts smoother.

Advantages

"Many hands make light work" used to be a common saying. It's still true. Dividing the manual writing according to systems or processes inherent in the product or according to special areas of writer expertise allows you to make the best use of writer talent and to get the job done more quickly and accurately.

For example, the writer whose specialty is filtration systems, calibration, or electrical systems will find it easier to write about that area than the writer who has to keep many different kinds of processes or procedures in mind.

Situations also arise in which a machine or product has used standard mechanical or chemical processes and is then suddenly altered by new technology or by the addition of an electronic component. These kinds of alterations have frequently occurred, for example, in products involving computers, numerical control, and robotics (e.g., devices for welding, spraying, assembly procedures, systems integration, and quality control). In such cases, the best use of talent may be to ask the technician or engineer–designer who created the new component to write the segment describing its function. An editor can then review the segment to make sure the style is user friendly and not filled with technical jargon.

Disadvantages

"The camel is a horse designed by a committee" is another saying, and one that applies here. Too often, the team-written manual has camel-like lumps and bumps. Such manuals move by fits and starts from one segment to another. They sometimes have ill-matched writing styles and formats.

Users find these manuals very difficult because of their redundancy, lack of cross-referencing, and chaotic organization. In brief, the chief difficulty with the team-written manual is the coordination of several writers' work into a smooth manual that looks and sounds as if one person had written it.

Coordinating the Team Effort

The team-written manual is a reflection of company structures and procedures, as well as management styles and individual personalities. Coordination of team writing should make the best use of available time, talent, and money. The question that needs to be answered up front is "Who decides what and when?"

Team-written manuals often have a team leader, a manager, or a service publications editor who has the final responsibility for the finished manual and a unified concept of what the finished manual will look like. That unified concept may be the joint creation of the writing team, However, once the conceptual framework is established, leaders are often responsible for the scheduling, assignment of manual segments, creation of clear instructions for what each manual segment is to include, and final coordination and editing of the completed manual.

Team leaders should have strong writing and editing skills because they will have the job of making language, style, and format internally consistent. A good team leader will make use of instructions, writer guidelines, writer checklists, regular (even if informal) meetings or check-ins — any procedure that helps writers know what is expected of them and when.

WHAT THE WRITER NEEDS (BESIDES MORE MONEY)

Writers, whether working alone or in groups, in small or large companies, have some very basic needs. To do a good job, they must have

- *T ime*
- *Access to information*

Time: How to Schedule It

Our comments on time as a basic writer's tool are directed especially to managers. Technical writing is deadline writing. More errors and slapdash jobs can be explained by time pressures than by incompetence. Managers often need to be reminded that writing takes time. Writers themselves usually do not have to be convinced.

Technical writers know that writing takes more time than anyone would ever guess. You're only putting a bunch of words on a page, right? So what's the big deal? Even writers, however, often underestimate how much time anything takes. We have surveyed writers in our seminars who estimate that even an average one- to two-page business letter or memo may take several hours to compose. Now add up the number of pages in a manual. See?

The good news is that once writers and/or managers have set up effective, consistent ways to collect information and have gone through the manual production process at least once, subsequent manual production proceeds more quickly.

However, totally new products need especially generous lead time for creating the manual. Some of the vital information may not be available until the last minute, for example, when the prototype is completed and tested. Ideally, you should always build in extra time to track down and talk to all the information experts involved, and to accommodate those guaranteed-to-appear last-minute changes.

We can't overemphasize this fact of writing life: *Manual writing is deadline writing*. Writers who have worked in any way on at least one manual already know this. Show this section to your boss. Again. Managers need to be reminded of this fact when they ask the impossible of writers. They'll still ask, but maybe with a bit more sympathy and a few more hours.

It would be truly wonderful if, when a company planned to develop a new product, someone came to the technical publications department and asked, "How long will you need to develop the appropriate documentation for this project?" And then doubled the answer. Instead, marketing sets the shipping date in consultation with engineering and manufacturing...and technical publications is expected to have the manual ready to ship with the product. However, writing takes time. How can you develop the manual in parallel with the product?

Manual production becomes much easier if, at the planning stage, you can determine approximately how much time you can devote to each part of the production process. Ideally, a product is rarely allowed to leave the production facility until the accompanying manuals or instructions are complete and ready for distribution or packaging with the product. The controlling date, then, is the deadline when the product is scheduled to be shipped, sold, or delivered. Establish this final deadline and create a work-flow schedule that allots time for the following phases.

Phase 1: Initial planning

- Clarifying the manual's functions
- Analyzing users
- Developing outlines and/or storyboards
- Setting writers' assignments
- Collecting information
- Creating writer checklists and guidelines

Phase 2: Preparing the manual

- Writing and layout
- Creating visuals
- Reviewing
- Editing
- Revising
- Preparing final copy
- Printing

Tasks in Manual Production: Initial Planning

The following sections describe the activities involved in each of the tasks of manual production. The sequencing and overlapping of the activities listed above vary enormously from company to company. Here is a brief look at what is typically involved in each of the tasks.

Clarifying the Manual's Function — Manuals are seldom stand-alone documents. They are usually part of an array or a family of supporting documentation for the product. Think about how the manual fits into the documentation family. If you think about the manual as part of the cluster of documentation, you are more likely to have complete coverage of all the essential areas as well as greatly reduce later headaches arising from overlap and repetition.

Analyzing Users — Steps for user analysis are described in Chapter 2. Analyze the user before writing the manual — this will allow you to decide on appropriate format and language levels for the manual. User questions may be employed to develop the outline and to determine what major sections the manual should contain.

Developing Outlines and/or Storyboards — Develop outlines or storyboards for major sections and/or chapters of the manual. Again, employ user questions to help you establish major sections of the manual (see Chapter 2). These outlines should be detailed enough to help you refine cross-referencing and to prevent redundancy and overlap. Good outlines let you see major blocks and chunks of information before you commit yourself to the actual writing. These chunks or modules are much easier to move around and refine at the outline stage than they are after you start writing actual text.

Setting Writers' Assignments — These may be for a complete manual or for segments of a manual. Decide whether to use solo writers or writing teams and assign responsibilities for editing the manual and for making final judgments on the physical look of the manual, including binding, paper stock, and page size.

Collecting Information — Sources of information may be customers, product designers, engineers, and personnel in marketing, sales, research and development, product safety, field service, and production. Other sources may include product histories, files, photographs, test data, and reusable materials from other manuals. Information gathering goes on from start to finish in manual production.

Creating Writer Checklists and Guidelines — These may be prepared by the writer(s), editors, and/or supervisors. Guidelines include lists of specifications and dimensions, special safety hazards, and new design features of the product. These may also include standard in-house instructions for format, standardized glossaries of terms for parts and procedures, and the types and sizes of photographs and drawings that may be used.

Tasks in Manual Production: Preparing the Manual

The rest of this book deals in considerable detail with such activities as writing and layout, visuals, and the editing and revision that make up final design of the manual. Preparation of final copy and printing choices depends, of course, on whether you are publishing in-house or by outside contractor.

Phases of Manual Production

Manual writing is seldom a series of smooth steps. Activities overlap and often occur simultaneously. Those activities most often squeezed out or omitted are clarification of manual function, outline development, and user analysis. Time spent on these is not time wasted but time gained, because it sets up the workplace and establishes a clear picture of manual function.

Phases of activity will vary in their sequencing, depending on the size of the company and its organizational structure. For example, some companies have not one, but several review or editing steps that may come early or late in the work-flow schedule. Further, the creation of visuals, the planning of format, and the writing of text usually occur simultaneously. Finally, the best of schedules will develop glitches — missing data, late photographs, key people sick or out of town, and/or last-minute design changes.

In short, you need to sequence and overlap the steps we suggest in a way that makes sense for your company, but do not omit Phase 1.

Scheduling Responsibility

In large companies, manual scheduling is usually handled by team leaders or publications editors. If you are working as a solo writer, you will have to do much of the scheduling yourself. Schedule yourself extra time when you have to collect information from other people or when you have arranged for outside help (e.g., with typing, drawing, or printing). Things always take longer than you think — especially with manual production.

Information: How To Get It

Odd as it may first seem, we have met many writers whose chief frustration was a lack of information about the product. Engineers are out of town troubleshooting another installation. Technicians can't hook the system up because they're working on another project. Designers are too busy to explain the basics.

Writers may ask to see the product and be refused. They may ask for scheduled time to review the product with designers, technicians, engineers, or safety personnel and be told that there is no time. They may ask for a working model, a prototype, or at least a photograph and get a flat "no" for an answer — or they may be housed in an office miles away from where the product is produced.

Admittedly, many people may be clamoring for a prototype of a new product. Marketing wants it, engineering is working on it, and product safety needs it. When the pressure is on and deadlines must be met, writers often come last. However, if the manual is to perform its function, writers must have information and management must provide the procedures to help them obtain it. Information gathering is an important first, and ongoing, step in planning the manual.

People tend to think of writing as a solitary occupation — the writer alone with resource material and a computer, hammering out paragraphs of golden prose. The actual writing may be solitary, but for technical writers, the resource material is almost always other people. A good manual writer spends much of the workday out and about, talking to people — engineers, technicians, service experts, even customers — because those are the people who have the information. However, these people are all busy with their own jobs. How can you get them to take time to help you?

Tactics

In a very small company, the writer may need only to lean across the desk and ask a co-worker for information. In very large companies, an e-mail might need to go halfway around the world. Successful writers we have talked to mention three key tactics for coping with obstacles. Note that we did not say "solving these problems" — they will never be totally solved. Writing manuals is like shooting at a moving target: products constantly change to incorporate innovation or to meet changing market needs. The manuals that accompany those products must also change. These three tactics, however, will help reduce the chaos.

1. Make Yourself Part of the Product Development Team — If your company practices concurrent engineering, it will understand this concept a little better. The earlier you can be involved in the product development effort, the earlier you can start writing the manual and the more timely information you will have. If your company does not already have the technical publications department as part of the product development team, volunteer. As one technical writer put it, "I just started going to meetings to which I wasn't invited. After a while, they expected me to be there!"

Also, show up at hazard analysis/safety meetings. Knowledge of safety concerns is vital, and your "outsider" position may help others who are more closely focused on creating the physical product see the design from the user's point of view.

Basically, the manual and the product should be designed in sync. Try to make it so in your company.

2. Cultivate Contacts in Key Areas (and Do a Few Favors) — Develop relationships with one or two people in engineering, service, and so on. If you cultivate a good working relationship with people, they will be more likely to pick up the phone and call you to tell you about a design change rather than to wait passively until you come to them.

Developing these contacts may mean trading favors. As the writing expert, you may be asked to edit letters or look at documents that are outside your job description. When you can, be friendly and helpful. You'll be in a better position to ask for help when you need it.

3. Develop the Manual As a Series of Modules — One of the writing techniques we have found to be helpful is modular writing, in which the manual is conceived of as a series of nearly self-contained segments. Some writers go so far as to limit the size of these segments to no more than a two-page spread, since that is all the reader can see at one time with book-style manuals.

However you define a module, the principle is the same: break the writing task down into bite-size units. This way, you can work on the different sections independently of one another. Even if you don't have all the information for Section A, for example, you can still work on Section B. In addition, the modular writing technique makes it easier to revise manuals (you can substitute new modules for those affected by model changes and leave others untouched) and to use relevant modules in more than one manual.

These suggestions will help you keep your job manageable. Adapt your information-gathering systems to the realities of your company and be on the lookout for ways to help the right people talk to each other. In later chapters we offer other help for dealing with the expected chaos of manual production.

Listed below are some of the successful techniques used by companies to improve information gathering.

Help in Gathering Information

- Product development meetings. These meetings are one of the richest sources of information for manual writers. When writers participate, from earliest phase to final form, in these meetings, they have a steady flow of information that gets plugged into the manual.
- Product safety committee. This group is a good source of information about key safety features and messages. If it is a hazard analysis committee, it can give the writer a more comprehensive understanding of the safety concerns with the product.
- Placement of writers' offices/desks near production, research, and test facilities. This location assures that writers are not working in a vacuum. One look at a product is worth 20 phone calls.
- Orderly file system. This system allows the writer to call up and reuse materials and modules prepared for other manuals, especially if the manual describes a new product with only minor design changes or a slight model change.
- Product history. A complete product history alerts writers to places where bad manual writing may have caused operator problems, accidents, or death. This product history may live with the product liability/hazard analysis team and may be stored in their safety library.
- Scheduled walk around. A viewing of the product or prototype well in advance of the manual deadline helps the writer immensely in understanding the equipment and the manual.
- Style guides. Writer guidelines and style handbooks prepared in-house by editors (or the writers themselves) save time and ensure a consistent manual.

- Writer's checklist. This information is provided by engineering, marketing, sales, product safety, and/or testing — whatever group bears responsibility or has the information. The checklist should include
  - Specifications and dimensions
  - Brief description of product function and use
  - Important safety features and hazards
  - New design features

All of the suggestions listed above are subsumed under a single prerequisite: *writers must have access to information.*

SUMMARY

Creating an effective manual needs intelligent allocation of time, money, personnel, and organizational support. Try to help your company see that the technical publications department is the information hub of the company, and as such deserves support.

Know where you as a writer/publications department fit within the company and play from the strengths of that location. Know how to work independently, but keep up a network inside and outside the company. Learn how to thrive, not just survive, in chaos.

The entire company, not just the writer, needs to remember that manual writing is deadline writing. A company gets a better manual if the writer gets both time and information. One way a writer can begin to get those is to be a part of the product development team. In fact, developing the publication concurrently with the product brings out the best of both.

If deadlines get too tight and resources scarce, ask your company managers, "Do you want the manual (1) good, (2) fast, (3) cheap?"

They can pick two.

REFERENCES

Bryan, Mark, with Julia Cameron and Catherine Allen. *The Artist's Way at Work*, William Morrow, New York, 1998.

Cameron, Julia. *The Artist's Way*, Jeremy P. Tarcher, New York, 1992.

Cooper, Alan. *The Inmates Are Running the Asylum: Why High-Tech Products Drive Us Crazy and How to Restore the Sanity,* SAMS, a division of Macmillan Computer Publishing, Indianapolis, 1999.

Galosich, Allison. Operation HAACP, *The National Provisioner,* January, 1999.

Norman, Donald A. *The Design of Everyday Things*. Doubleday/Currency, New York, 1990 (originally published as *The Psychology of Everyday Things*, Basic Books, New York, 1988).

Perry, Susan K. *Writing in Flow*, Writer's Digest Books, Cincinnati, 1999.

Schriver, Karen A. *Dynamics in Document Design*, John Wiley & Sons, New York, 1997.

Wishbow, Nina. Home Sweet Home: Where Do Technical Communication Departments Belong? *Journal of Computer Documentation*, vol. 23, no. 1, February, 1999.

CHAPTER 2

Analyzing the Manual Users

OVERVIEW

User analysis is one of the major tasks in planning and producing manuals. Writers need to have a clear picture of manual users and their needs. What the users need to know, and how they'll use that information, should guide manual writers in choices of arrangement of manual sections, integration of visuals and text, language level, and safety warnings. The guiding principle and fundamental responsibility of writers is to design manuals first for the customers — not for the company and not for the courts. Keeping the customer primary, as we have explained elsewhere, actually ends up pleasing courts and companies, too.

This chapter gives guidelines for user analysis and user feedback. We discuss why and how users read manuals, who users of your products really are and how to find out, and how to design one manual to please many different users.

We encourage you to begin with what you don't know about your users. We offer lists of questions for you to consider as you design your manuals. We show how questions typically asked by users can help you create manual outlines and decide on major sections of the manual. We suggest techniques for collecting user feedback and for using that feedback to revise and update manuals.

Overall, we help you simulate the person-on-the-street user to avoid your own "shop blindness" — the inability to see your products as a first-time user might. At the end of the chapter, you'll find a checklist to help you analyze your real users and see your products the way they do.

DO PEOPLE REALLY USE MANUALS? YES, BUT...

We've all told the stories: Customers calling service support with questions the manual clearly answers on page 3. Former friends and ex-relatives tossing our shrink-wrapped labors with a cheerful, "Oh, I don't bother with instructions." VCRs across four time zones blinking 12:00 in eternal testimony to our communal failure to read the manual.

These stories tell us a couple things, perhaps: that what's clear to you is not clear to everyone, and that people close to you can be cruel. What they *don't* tell us is that people never, never, read manuals — because they do.

Recent studies "suggest that the facile assumption that no one reads documents is simply wrong. If anything, … users are telling companies that they want better documents and that they contribute significantly to customer satisfaction" (Schriver 214). Document design expert Karen Schriver found that only 4% of people interviewed during a use-of-product study said they never used the manual — 96% said they did. "Seventy-nine percent of users report they would buy from a company they thought had clear communications," and "more than half of participants would consider paying more for a product if they knew it had a clear manual" (Schriver 223). Customers are also making headlines by returning products where poorly written manuals directly contribute to confusion and misuse (cited in Schriver 240).

What we've found from others and from our own work with customers and companies is that people do use manuals — they just don't use them as we would expect, or, frankly, would like. In other words, the "failure to use the manual" in many cases is more accurately "failure to use the manual as the technical writer expects it to be used." The failure is ours as writers, then, and it is a failure of our assumptions.

If we stop making assumptions about how users *should* read our manuals and look at how they actually do use them, here's some of what we find.

- People will use manuals only if the output meets or exceeds the effort. Do they get something for their time?
- People read with memory — their past encounters with instructions shape their present experience.
- People almost always would rather ask a person than read a text. People read only if they can't get information another, easier way, and they read as little as possible.
- People read a manual while they are doing something else — working with the product. This simultaneous activity is crucial to understanding how to design a manual to help a user.
- Many, many people are reading to use a product on the job. They are not reading for fun; they are reading only to use a product to get something else done. A manual, and a product, for that matter, should be a service, not a hardship, for the user.
- People are often in a hurry, under stress, anxious, or at least uncertain when they pick up a manual, particularly if the product is unfamiliar to them or appears to be technically sophisticated — or involves software of any kind.
- People read different manuals in different ways depending on the products. If they think the product is simple and shouldn't need (or they don't want there to be) instructions, they may not read at all.
- People with experience in reading manuals are opportunistic readers — they will pick and choose what to read and, equally important, what not to read (see Schriver 165).
- People do not *deliberately* set out to read poorly or use the product ineptly or unsafely, as Lila Laux, human factors expert with U.S. West Technologies, points out (Laux).

WHAT DO USERS REALLY WANT? WHY CAN'T THEY GET IT? AND SHOULD WE CARE?

Users want clear information that will help them solve their problems or answer their questions as they use a product. They also want to save face. Interview after interview with people during usability testing reveals this fact: people feel that a great many products, particularly software-based ones, and their accompanying manuals, make them feel stupid.

"Not feel stupid" is one of users' "personal goals" for using a product/manual, along with "not make mistakes; get an adequate amount of work done; and have fun (or at least not be too bored)" (Cooper 156). These "personal goals are always true and operate to varying extents for everyone. Personal goals always take precedence over any other goal, although — precisely because they are personal — they are rarely discussed," argues Alan Cooper, an interaction design consultant and one of the creators of Microsoft programs. "When software makes users feel stupid, their self-esteem droops and their effectiveness plummets, regardless of their other goals. Any system that violates personal goals will ultimately fail, regardless of how well it achieves other goals" (Cooper 156).

When users can't figure out a product, when they feel stupid, they blame themselves. In some cases they may be right, but not always, and not as often as they seem to believe. Schriver's analysis of people's responses as they actually use a variety of products found that, female or male, old or young, "users blamed themselves for the problems they experienced more than half the time." (They blame the manual only 30% of the time.) Schriver finds these patterns of self-blame particularly disturbing "because of their potential cumulative effects. Over time, people's repeated experiences with badly designed products and instruction guides may convince them that they're incompetent as both readers and users of technology" (Schriver 222).

Cooper believes needlessly complicated products are at fault. Why? With computers, for example, programmers tend to be the ones who design the final form, and they design with the computer in mind, not the human user. Not designing for the user, Cooper says, is already creating a kind of "'software apartheid,' where otherwise normal people are forbidden from entering the job market and participating in society because they cannot use computers effectively. In our enlightened society, social activists are working hard to break down race and class barriers while technologists are hard at work inadvertently erecting new, bigger ones. By purposefully designing our software-based products to be more human and forgiving, we can automatically make them more inclusive, more class- and color-blind" (Cooper 11).

Manuals often compensate for problems ill-designed products create for the users. In a way, good manuals buffer the harsh effects of poorly designed products (and poor manuals compound them). As our technological abilities and our consumer demands increase, and as our product development time often decreases, we experience a growing number of these effects, among them:

- **Creeping featurism.** "The tendency to add to the number of features that a device can do, often extending the number beyond all reason" (Norman

173); also called the "Akihabara syndrome" — "named after Tokyo's famous consumer electronics district" (cited in Schriver 239).
- **Cognitive friction.** "The resistance encountered by a human intellect when it engages with a complex system of rules that change as the problem permutes" (Cooper 19–20); a lawn mower or a Swiss army knife is low in cognitive friction, a cell phone or a personal computer is high.
- **Productivity paradox.** "New technology that is designed to make us work faster and smarter, but actually only makes us work slower and with less confidence" (Schriver 222); this paradox is perhaps explained by cognitive friction, and in turn explains why many people would rather stick to the old product which they basically know how to use than switch to the newer model which they'd then have to relearn.
- **Illusion of knowing.** "A reader's belief that comprehension has gone smoothly when comprehension has actually failed" (cited in Schriver 226); students quite frequently engage in this illusion after exams; and although this problem is the user's, not the product's, it still becomes the writer's to counter.

Manual writing would be a lot easier — and in some cases totally unnecessary — if products were designed better from the beginning, i.e., with users in mind. Creeping featurism and cognitive friction, in particular, are the effects of ignoring users' real needs.

Add these long-term effects to the shorter-term ones of company image and product liability (discussed in Chapter 1) and we see that designing for the user does indeed matter — to the customer, to the company, and to the writer. The question is not "Can we afford to do good user analysis?" but "Can we afford not to?" The answer is no.

WHO ARE YOUR USERS? HOW DO YOU KNOW? HOW DO YOU FIND OUT?

To work effectively, a piece of writing, just like a piece of equipment or a product, must be designed with three questions in mind:

1. Who is supposed to use it?
2. What is it supposed to do?
3. Where is it supposed to be used?

Until writers answer these questions, they cannot choose and arrange their material effectively.

The answers to the first question — "Who is supposed to use it?" — determine many of the answers to the second question, "What is it supposed to do?" For example, if the user of a manual is a first-time buyer of a riding mower, the manual writer must not assume that the user will know how to maneuver the mower on a hill. On the other hand, if the user of an electrical service manual is an experienced electrician, the writer can assume that the user will have a basic knowledge of electrical circuits.

The answers to the third question — "Where is it supposed to be used?" — take into account what Laux calls "the environment of use: the real circumstances in which your product is used." The environment is both physical and psychological: is it an office or a field, dry and cool or wet and humid, quiet or noisy, clean or dirty, dark or well lit? Does the user have time to read at length, or must the user dive in and get a machine up and running? Is the user reading the manual at home, for personal use, or at work, for business? Will the manual be used on the shop floor, next to oily machinery, or in the office, next to the computer screen?

All these answers help the writer design, write, organize, and physically create a document matched to user needs. (No matter the answers, don't forget that most users will be doing at least two things at once: reading the manual and trying out the instruction step on the product.) Chapter 6 discusses how vital this question is for the design and placement of safety warnings on equipment/products.

Many companies today are not only trying to figure out who is really using their product, but what their core product really is. A company that sold its product piece by piece now finds itself designing integrated systems of pieces. Another company now markets its technical know-how as much as it does its hardware. Customer demands change, as do the customers themselves. The initial customer for a technically sophisticated device that measures oil in pipelines might be a highly educated company engineer, but the actual end user might be a line operator who never finished high school.

Find out all you can about the real user of your product. Knowing user behavior is particularly important for product liability prevention. Laux suggests a number of ways to find out about your users:

- Sales data and field service reports
- Observations and reply cards
- Studies (such as failure modes analysis for product liability)
- Reports and surveys from consumer, manufacturer, and industry groups, unions, safety councils and commissions, court cases (Laux 33)

Listen to customer complaints, questions, and service requests. If you manufacture equipment, another way to find out how people are using your product is to track parts ordered. The same part ordered again and again, for example, tells you something about the quality of your product or the (mis)use of it by the customer. Repeated service calls also give clues about user behavior.

For each manual you write, and for each manual update, ask, "Do you know your users? They may not be who you think they are. They may not be who you intend them to be" (Laux 31).

SPECTRUM OF USERS

For many products, the wide range of potential users poses a problem for manual writers. At one end of the spectrum are the professionals; at the other is the general public. (See Cooper 182–185 for a different spectrum for software users. He says that programmers design for experts, marketing designs for beginners, and that everyone ignores the largest and most consistent group of users, "the perpetual intermediates.")

As a result, when you first ask yourself, "Who is supposed to use this manual?" you may only envision an undefined, anonymous buyer. Let's take a look at some definitions of professionals and general public and at some user characteristics and distinctions. These guides should help you to bring into sharper focus the hazy picture of your manual user.

Professionals

Certain kinds of manuals and instructions are written expressly for professionals, e.g., the service manual for the trained mechanic or the instructions for a surgical probe. In these cases, a professional may be defined as a user who is likely to have special or in-depth knowledge of the product, its purpose, and the technical terms used to describe it.

For example, some manuals are written exclusively for such professionals as these:

- Licensed electricians or plumbers
- Master carpenters or mechanics
- Trained and licensed service and repair persons
- Registered engineers (industrial, civil, mechanical, electrical, computer, nuclear, chemical, aeronautical, metallurgical)
- Medical professionals (nurses, doctors, lab technicians, physician assistants, nurse practitioners, sports medicine trainers)
- Computer professionals (hardware or software programmers and designers, computer service and maintenance personnel)

Manuals for the professional are often easier to write because you can make certain safe assumptions about the user's understanding of the product. That confidence is usually reflected in the level of language you are free to use in talking to the professional. Even with professionals, however, avoid unnecessary jargon and excessive technicality.

Quite often, the same product will have two sets of users. In that case, the product will have two manuals, written at different language levels, or one manual with separate manual sections, some written for professionals and some written for general public users. In the second case, each section must be clearly labeled so users know whether to read or skip.

Example 2.1 contains two passages that describe the same mechanism. The writer has altered the style, however, to meet specific users' needs. These two passages could both occur in the same manual — the first in a section directed to surgeons or trained technical personnel, and the second in a section directed to equipment assistants and nontechnical personnel. (In Chapter 7 we discuss service manuals.

Example 2.1 Changing style for different users. (Adapted from Cooper Vision Systems *Operators Manual, Model 8000,* Cavitron Surgical Systems, Irvine, CA. With permission.)

Passage A — For the Professional

A peristaltic pump is used to create the suction for vacuum. You can use different vacuum levels for the various handpieces and "tips" employed during surgery.

A constant volume peristaltic pump delivers a constant flow rate of 28 ± 2 cc/min. The pump is driven by a regulated DC control voltage that has less than 2% output variation of 106 to 128 VRMS and a load variation of 0 to 75 oz-in. of torque.

Passage B — For the General Public User

Venting action causes the vacuum level at the port to drop to less than 2 inches of water in less than 300 milliseconds and occurs automatically each time the foot switch moves from Position 2 to Position 3.

An important part of the vacuum system is the vent control. A simple way to understand venting is to think about what happens to fluid in a soda straw. The fluid is sucked up into the straw. (This is analogous to the vacuum buildup taking place in the ABC system.) If you place your finger over the top of the straw, the vacuum is maintained and the fluid stays in the straw with little or no leakage. To release the fluid, remove your finger from the top of the straw and allow air to enter. In essence, this is what happens in the vacuum system of the ABC machine.

Service manuals for professionals rely on the user's in-depth knowledge of the product, but also demand that the manual be put together somewhat differently than a manual for the general public.)

Both passages describe the same mechanism. Passage A assumes the reader knows the basics of pump operation and venting, as well as terms such as "output variation," "load variation," and "torque" without having them defined. Passage B uses a minimum of technical language and explains the venting action by means of the simple analogy of the soda straw. Manuals written for a more general audience often use comparisons with common items or activities to help clarify technical processes. Manuals for professionals may use more technical language and visuals, but must still be as clear and logical as manuals for the general public.

General Public

Far more common are manuals for the general public, sometimes called consumer manuals. The term *general public* seems at first glance to be so broad that it defies definition. One way to understand it is by looking at the categories of products usually intended for the general public — consumer products:

- Appliances
- Automotive products
- Biomedical devices (blood pressure kits, contraceptives, prostheses, heating pads, braces, contact lenses, glasses, hearing aids, dentures)
- Construction equipment
- Drugs and health products

- Farm and industrial equipment
- Firearms
- Foods
- Household products (soaps, polishes, cleansers, sprays, pesticides, step stools, ladders)
- Office equipment
- Paints, general-purpose chemicals (fertilizers, solvents, removers)
- Power and hand tools
- Sporting goods (bicycles, skis, swimming pools, skateboards)
- Toys

We can also look at the general public user or consumer in relation to biology, literacy and language fluency, and technology/technical sophistication.

Biology

A general public user can be male or female, a child to an elderly person. If you write instructions for skis designed for the resilient 16-year-old body, you have no assurance that a 50-year-old won't try them. If you make a caustic toilet bowl cleanser for an adult to use, you have no assurance that a preschooler, who can't read the label, won't think it's just another colored powder or drink. Cultural customs change, too. How many products once marketed solely for one gender, for example, are being used by everyone: blenders, power saws, hair dryers, and shotguns?

Literacy

Illiteracy and declining literacy are modern realities. Instructions for use by the general public should not assume literacy. If the product is especially complex, unusual, or hazardous, reliance on visuals or pictures is essential in the manual and in instructions affixed to the product. (Chapter 6 discusses the special problem of safety warnings.)

Language Fluency

The number of people in the workforce for whom English is not their mother tongue is growing. These people may speak two, three, or seven languages; English just might not be the one they're fluent in yet. If you know your product users are bi- or multilingual, use lots of visuals and write in another language in addition to English. (Chapter 8 discusses the specific issue of writing for international markets and complying with multinational directives and regulations.)

Technology/Technical Sophistication

The general public user may be technically sophisticated, technically naive, or somewhere in-between, like Cooper's "perpetual intermediate" software user. If your product is intended for general use, aim the manual at the beginning and intermediate users. Most people try to move out of the beginner stage as quickly as possible, then tend to hang in the intermediate stage indefinitely, learning the least amount they need to know to do what they want. As Cooper points out, this description is particularly true of software users (Cooper 182–185).

Technically sophisticated users will either skim these sections or jump over them entirely. If you aim the manual at the technically sophisticated — and many manual writers, particularly if they themselves are technically adept, do — you will lose the more inexperienced user. The manual is then useless to that reader and goes unread.

Other Characteristics of Users

Most manufacturers, by knowing the market for their products, can aim or target manuals, written instructions, and warnings by using the following list of user characteristics. (For other ways of analyzing user characteristics, see Cooper's Goal-Directed Design® tools 123, and Coe's user profiles 198.) The list gives you more information for deciding if your users are professional or general public users.

Personal Characteristics of Users

- Do they use this machine or product almost every day or only once in a while?
- Are they likely to have used other products like it?
- Do they do their own routine maintenance?
- Do they do their own repairs? Should they?
- Do they understand technical language?
- Do they understand charts, circuit diagrams, mechanical drawings?

Conditions of Manual Use

- Will they use the manual only to learn how to set up, use, or operate the product?
- Will they refer to the manual or instructions often?
- Will they use the manual or instructions only if something breaks, fails, or appears to be abnormal?
- Will they read the entire manual or only a section here and there?
- Will they be able to look at the machine or product when they are reading the manual?
- Will the illumination be good?
- What are other physical conditions of use (wet, dry, oily, noisy, dusty, cramped, etc.)?

Information Wanted

- Basic instructions for use, operation, and adjustment?
- Routine maintenance procedures?
- Sophisticated service procedures?
- Specifications and parts lists?

- Troubleshooting procedures?
- Explanations of new technologies or product features?

Remember that the same user may respond differently to different products. A neurosurgeon may be technically sophisticated with operating room equipment and an inexperienced beginner with power paint sprayers. A professional in one situation can easily be a novice in another.

Consider how the following situations demand simplicity and nontechnical clarity in written instructions.

- The first-time homeowner repairing a faucet? Changing the air conditioner filter? Rewiring the kitchen outlet? Building a deck?
- The independent auto mechanic with 30 years of experience servicing an older car? Servicing a late-model car with parts formerly mechanically controlled, now computer controlled?
- A 12-year-old learning to use her computer? An 80-year-old learning to use his computer? Both of them playing a video game? Either of them programming a VCR?
- A maintenance supervisor of heating and air-conditioning equipment fixing an air conditioner in a 40-year-old plant? Maintaining new equipment that uses computer technology and automated controls?

General public manuals should be simple, clear, and nontechnical. Professional user manuals should be simple, clear, and as jargon-free as possible. Whenever the spectrum of users could range from sophisticated to inexperienced or naive, make manuals as nontechnical as possible. Remember that even the most advanced and knowledgeable user appreciates clear and simple instructions.

USERS' QUESTIONS AS MANUAL ORGANIZERS

After you have established a clearer idea of who is supposed to use your manual and under what conditions, you are ready to determine what the manual should contain and how its major sections should be organized. Ask yourself, "What is the manual supposed to do?"

Your users can be your guides here. Lay out the manual sections as if they were responses to users' questions. To do this you must step back from your product and try to see it as first-time buyers or users might. What questions would they ask?

- The computer keyboard looks a lot like my typewriter. What's different about it? How do I underline? Indent? Set the line spacing? What else can it do a typewriter can't?
- My boss tells me this new paper-making machine is state-of-the-art equipment. It looks like our old one to me. What's new? How much retraining will my line operators need to keep it running right?

- Interesting-looking toy — how does it work? What does it do? How do I make it stop?
- I heard about this product on TV. Let's see what's in it to get out puppy stains on the carpet. Do I have to mix it with water?
- What are the installation-space dimensions for this? Must it be fireproofed?
- Can I use my hair dryer in the bathtub?
- Can I turn off the printer before I shut down the computer?
- If we buy this pump, what kind of retrofit do we need on our old equipment?
- How do you change the settings on the copy machine? How do I set it to copy and collate?

Imagine the buyers or users of your product talking to your industry vendors or shopping in a store or a dealer showroom. Buyers look at the product, talk to salespeople and vendors, or read the sales literature and instruction manuals. They do this because they have *questions*. Answers to those questions can form the major sections of your manual.

Look at the chart to see how a user question can form a section of your manual. The column on the left gives you a typical user question and its corresponding manual section. The columns to the right contain typical answers for two products, an exterior paint and a heavy-duty wrecker. These answers could be used to make up the major sections of the manual.

Chart Showing Manual Sections Derived from User Questions

Manual Section Derived from User Question	Answer to User Question about the Product	
	Exterior Paint	Heavy-Duty Wrecker
Scope (What is the main function of this product?)	One of 25 products for protective exterior coating	Mounted on truck chassis; used for towing, lifting heavy vehicles
Description of product (Is this what I'm looking for? Introduce me to it.)	Oil base, high quality, for residential or commercial buildings; 15 colors available	Boom and two winches; two telescoping outriggers on upright mast
Theory of operation or intended use (How does it work? What is it for?)	Protective coating for wood, asbestos, brick, stucco	Recovery operations using winches; choice of pulling or towing by front or rear wheels
Special feature or design details (What is special about it?)	Durability, nonfade color, controlled replacement color formula	Boom ratings: extended, 12 tons; retracted, 35 tons; winch ratings: safe load — 17.5 tons
Limits of operation or use (What are its limitations?)	Not for metal, glass, or plastic; brush application only — not spray; mix only with organic solvents — no water	Ratings apply only if truck chassis is adequate, both winches are attached to load, boom is at 15° from horizontal, load is lifted vertically
Setting up/turning on (How can I assemble it? Turn it on?)	Brush application; temperatures above 50°C	Wrecker installation on truck chassis requires special training

(continued)

Chart Showing Manual Sections Derived from User Questions (continued)

Manual Section Derived from User Question	Answer to User Question about the Product	
	Exterior Paint	**Heavy-Duty Wrecker**
Normal operation or use (What is normal use and life of product?)	Dries in 24 hr; 7-year life; can be washed	Heavily dependent on good maintenance and variations in weather conditions
Turning off/disposal (How do I turn it off? Dispose of it?)	Excess and accompanying solvents are flammable; precautions for handling	Turn-off controls
Abnormal operation (What tells me something has malfunctioned?)	Causes of cracking, peeling; not for internal use	Signs of malfunction; safety features; damage to cable, boom, or to load being towed or lifted
Preventive maintenance (How do I take care of it?)	Proper surface preparation; proper application; close lid tightly	Complex machine; separate repair and maintenance manual
Storage (How do I store it?)	Store upside-down; shelf life	—
Safety (How do I use it safely?)	Safety information will be found throughout the manual and on the product (see Chapter 6)	

USERS AS COCREATORS: FEEDBACK SOURCES

Usability testing is not just for the product. Usability testing is also for the manual. As we have shown, manual users are important as audiences and organizers of your manual. They can also work almost as active cocreators, rather than passive consumers, of the manual through user feedback. As soon as possible in the manual production process, talk to users. Visit customers' plants, if possible. Talk with your company's sales, manufacturing, installation, parts, and field service people. Listen to your customers and your co-workers.

User ideas and feedback are particularly helpful during initial design and then revision and follow-up stages of manual production. Writers who rely on user feedback tell us that this feedback is invaluable for debugging a manual before the final copy is printed and for assessing manual use and effectiveness after the product is sold.

Simulating the User: The Person on the Street

How do you do that simulation? We have found that company efforts to collect user feedback range from informal and occasional to formal and systematic. (See Coe 192–198 for other considerations with usability testing.) Here are some of the techniques companies use.

Informal and In-House

Some companies invite employees from other divisions, friends, family members — anyone who is a true stranger to the product — to "walk through" the manual, following its instructions and descriptions. Writers stand by and listen and watch, but they don't provide verbal backup to the manual unless the user asks for help. Wherever writers have to break in, explain more fully, or provide more information, the manual probably needs revision or clarification.

Beta Site Testing

Some companies, including those doing military contract work, have selected sites (beta sites) used for testing products and debugging manuals. At beta sites, users are asked to perform the operations described in the manual. For instance, an airplane mechanic may be asked to follow the manual for installation of a new landing gear. The mechanic is selected and identified as a typical user, someone trained to work on military aircraft. All difficulties and snags experienced by the user in following instructions are monitored and recorded. The manual is corrected, revised, and retested to assure that instructions are clear.

In the private sector, potential or long-time customers are asked to be beta sites for product and manual testing. Writers travel to the beta sites to observe the product and the prototype manuals being used by beta site employees. Companies who agree to be beta sites are usually compensated, by reduced price or some form of service. However, for many military contracts, beta site testing is mandatory.

Protocol Analysis

As people perform tasks using a manual, they are often doing several things at once — reading, using their hands and/or feet, and talking, either to themselves or to someone else. Protocol analysis watches what people do as they read and work, but it also listens. Sessions with users are videotaped, often from several angles. The manual users are asked to talk out loud, describing what they are doing or thinking. This verbalization of tasks is a rich source of information for manual producers. A user may stop, look puzzled, and ask, "What screw? Where is the slot for it?" or mutter, "Abort? Error? Retry? I did what this thing told me to. How do I get out of this? It keeps saying Error, Error. Help!"

Formal Systematic Analysis

Some companies dedicate considerable money and time to formal user analysis. The setups can be as simple as the use of a videotape (with audio recorder) to record the manual user at work. More elaborate user feedback setups place users at a desk equipped with a microphone, the manual, and videotape monitors. Manual pages are presented, and as users work their way through tasks, their activities and voices are recorded.

Design engineers, safety engineers, and technical publications staff monitor the users at work from a studio equipped with one-way glass. The videotape record of the user at work is digitally synchronized with a videotape of the manual pages. The combination of voice plus video of both user and manual page allows observers to know exactly what the users saw on the page, what task they were trying to perform, and what they said and did about it.

User Interviews and Surveys

Many companies now do follow-up surveys and field checks on their manuals by asking users for feedback. Some companies simply include a postage-paid card in their manuals asking users to assess manual effectiveness. Response in this way, however, is likely to be low unless the manual is very bad.

Some companies use manual hotlines, listing a toll-free phone number in the manual and inviting people to call if they have trouble using the manual. You can also invite e-mail response. Other companies make up a list of preferred, repeat customers and actively seek their feedback on manuals by making personal phone calls to customer–users or visits to sites where new products have been installed. These "personal touch" interviews, whether by phone or direct visit, are carefully planned surveys where specific questions are prepared ahead of time and feedback can then be used to improve manuals.

Benchmarking

Another way to get user feedback is to ask customers what they think about competitors' manuals and instructions. "Learn from the competition's successes," says technical communications consultant Marlana Coe. "Benchmark information whose presentation users dislike as well, so you can avoid the competition's failures" (Coe 268).

Making Manuals Match Users' Needs

The information you gather about your users will be your guide as you make choices in writing the manual. User analysis affects:

- Language level and reading level
- Choice and execution of visual and graphics
- Proportion of visual to verbal text
- Arrangement of segments of the manual
- Safety warnings
- Revision and updates

Again, particularly for general public users, lower the language level (some surveys find American adults read at a sixth or seventh grade level), increase the number of visuals, use white space to group information and make reading easier, organize simply and consistently, and repeat safety warnings more often the greater the danger.

Example 2.2 How user analysis affects language and reading level. Here are two different descriptions of the same procedure (removing front brake shoes from drum-type automotive brakes). (Passage A, from *Dodge Dart, Coronet and Charger Service Manual,* Dodge Division, Chrysler Motors Corporation, Detroit, MI, 1967, 5–7. With permission. Passage B, from Muir, J. and Gregg, T. *How to Keep Your Volkswagen Alive: A Manual of Step by Step Procedures for the Compleat Idiot,* John Muir Publications, Santa Fe, NM, 1974, 5–10. With permission.)

Passage A — From the Dodge Manual

1. Using Tool C-3785 remove secondary return spring then remove adjusting cable eye from anchor. (Note how secondary spring overlaps primary spring.)
2. Remove primary return spring.
3. Remove brake shoe retainer springs by inserting a small punch into center of spring and, while holding backing plate retainer clip, press in, and disconnect spring. Unhook cable from lever. Remove cable and cable guide.
4. Disconnect lever spring from lever and disengage from shoe web. Remove spring and lever.
5. Remove primary and secondary brake shoe assemblies and adjusting star wheel from support. Install wheel cylinder clamps (Tool C416) to hold pistons in cylinders.

Passage B — From the Muir Volkswagen Manual

Look at the brake assembly; you'll find there are return springs holding both ends of the shoes toward each other. Look at how they're fastened — see how they're anchored. You'll have to replace them the same way. Clamp the vice [*sic*] grips on the closest spring to you then pry it out of its hole with the screwdriver. Remove the other springs the same. Take out the two round springs with caps over them in the center of the brake shoe webs, holding the shoes to the brake plate. They'll come off with your fingers, so hold the pin in the back of one with your forefinger and push and twist on the little cap with the thumb and forefinger of the other hand. When the little cap is 90° around on the pin, it will come off and the whole thing will come apart. Take a close look, you have to put them back on .... Remove the other spring–cap–pin assembly from the other side. Now you can work them out of their slots. Snap a rubber band tight around the wheel cylinder slots so the wheel cylinders won't come apart.

You can manipulate these elements in a number of ways to match user needs and questions. Remember that there are always at least two equally good ways to say the same thing. Try out a couple variations, altering different elements of the manual. Subsequent chapters discuss these choices in greater detail. The two passages shown in Example 2.2 offer samples of how you can alter language levels to match the reading levels of general public and professional users.

The procedures described in these two passages are identical, but the writers had two very different kinds of users in mind. Strengths and weaknesses of the two approaches include the following:

Passage 1: Dodge Service Manual

- Is written for a professional mechanic
- Assumes familiarity with tools, parts, and mechanisms
- Is easy to follow with a numbered list format

Passage 2: Muir Volkswagen Manual

- Is written for a novice mechanic or do-it-yourselfer
- Anticipates trouble spots (especially those requiring unusual coordination of hand or hand and eye)
- Warns of disassembly problems
- Tries to use common terms for tools and parts
- Is harder to follow than the Dodge manual because of use of paragraph format instead of lists
- Is made wordy because of chatty style, but also more appropriate for a beginner

User Feedback to Fine-Tune Manuals: An Informal Survey

Feedback from users can also help you make adjustments in manuals by revising, updating, or altering manuals as successive models of a product come out. The following example shows how user feedback from a survey can sharpen insights on actual use of the manual. Talking with manual users, either informally or through structured surveys, can be informative and can give you insights on the strengths and weaknesses of your manuals.

To see what kind of feedback might be provided by manual users, we conducted a series of informal interviews with service people, dealers, and buyers/users of agricultural equipment. We tried to ask questions that would encourage people to talk honestly about areas of the manuals that most concerned them. We asked, "Do you use the manuals for your equipment? If so, how? If not, why not?"

We found that dealers and service people answer the questions somewhat differently than do buyers/users (farmers). Here are some of the responses to our questions. Notice, as you read, that dealers and service people stress the importance of the manual as a teaching tool and as a legal document. They are also keenly aware of different levels of technical sophistication among buyers/users.

The buyers/users — the farmers themselves — have occasional positive things to say, but have many complaints about manual effectiveness.

At the end of the user responses you will find our comments on what we learned from this simple feedback exercise.

Feedback from Dealers and Service People

"There's a big difference in some of these modern-day farmers. There'll always be some who think the manuals are too complex — they want to be told to tighten down a nut, not what its dimensions are. On the other had, if they're used to checking on a corn planter just by eyeballing it, and then they buy one with electronic equipment, they'll sit down and read every word."

"No use lecturing a farmer about stuff he's used all his life. He tends to think all engines are pretty much the same. He knows how to drive a tractor; he's been doing it all his life. He'll look at the manual if there's a new wrinkle, but he has to have it pointed out to him."

"I guess there's a lot of the 'good old American know-how' in most of my buyers. I get my calls when they've taken something apart or tried to fix it and can't. If it looks like they've botched a job and the labor bill is going to get bigger if they go on with it, they'll call us to come out to bring it in."

"Some of my buyers are buying for as many as 32 farms. They've got full-service departments to take care of their equipment — some of their mechanics are better than ours. They're very well educated, and they don't want to be talked down to."

"The ones that need the most elementary help are these 90-day-wonder suburbanite farmers with 15 acres of land. They're probably buying a small tractor for the first time — everything's new to them."

"You could mount the safety instructions in neon lights on the power takeoff, and 50 percent of the old-hand farmers will remove whatever gets in their way."

"It used to be that the operator manual got filed away or tossed somewhere. Now they're getting more and more use. The equipment is more complicated, more things can go wrong, more malfunctions are adjustment problems."

"Our service and delivery people use the manuals for teaching. In fact, we set up a school for all our buyers of corn planters, balers, and combines. The service people run those schools, and they show the buyers what kinds of adjustments and troubleshooting are likely to present problems. They use the manuals to show a farmer what he can do for himself before he calls for help."

"One of the biggest complaints is that the manuals don't go deep enough. The pictures are fuzzy, or they don't know what kind of tool they need."

"We think the manual is so important that we require dealers to register the serial number of the manual as well as the equipment. (1) We want to make sure he's got it and have proof of it. (2) We want to emphasize its importance to our customer. (3) Liability settlements are getting bigger all the time — there are more and more ways to get hurt."

Feedback from Buyers/Users — The Farmers

"I wish they'd always give you the manual that goes with your equipment. Some of them don't really match the machine you've got. I can't tell you how many times I've stared at those pictures and wondered if that funny little thing sticking out of the picture is the same as the funny little thing on my machine."

"The pictures are terrible. Half of them are too small and full of those damn little numbers. You can't see what the arrows are pointing at, and you have to keep flipping back and forth. Most farmers I know have eyesight that's none too good — all that crap flying around in the air, you know."

"You ought to talk to about 500 farmers. I'll bet they'd all tell you those things are put together by a bunch of engineers sitting in offices with fluorescent lights. Most farmers aren't engineers. They're working in mud or in a dark tool barn."

"I weld a lot. I know there's some stuff I shouldn't 'cause when I try to turn on something, I find out what I welded doesn't work so good anymore — wasn't made to be welded. But some of my stuff is pretty old, and a trip to the dealer is 35 miles and a 5-hour ordeal, and then I might not even get the part. Who's got that kind of time?"

Summary of Survey Feedback

Given a chance to talk about how they use manuals, these farmers, dealers, and service people have good and bad things to say about them.

From their responses we learn the following:

- Users are general public, but range from technically sophisticated to technically naive.
- Manuals for products with new design features or repeated service problems get more use.
- Confusing or simply too-small graphics are a problem for many users.
- Users sometimes are given the wrong manual or a generic manual intended to serve for several models of a product; users do not like generic manuals.
- Dealers and service people rely heavily on manuals and sometimes use them for teaching.
- Conditions for manual use are sometimes bad (field, mud, dark, barn).
- Farmers tend to use manuals for doing breakdown, maintenance, repair, and setup, or for understanding new design features.
- Farmers work under pressure and at a distance from help, and so do much of their own service and repair.
- Farmers probably seldom read the whole manual for a product they are familiar with (or think they are).

CHECKLIST: USER QUESTIONS AS ORGANIZERS OF THE MANUAL

This checklist can be used either to plan and outline a new manual or to evaluate a manual you have already written. (Adapted from Ranous. With permission.)

		User Questions	Segment of Your Manual That Answers Questions
1.	Scope	1.	1.
2.	Description	2.	2.
3.	Theory of operation	3.	3.
4.	Design detail	4.	4.
5.	Limits of operation	5.	5.
6.	Setting up and turning on	6.	6.
7.	Normal operations	7.	7.
8.	Turning off	8.	8.
9.	Abnormal operation (troubleshooting)	9.	9.
10.	Preventive maintenance	10.	10.
11.	Storage	11.	11.
12.	Safety*	12.	12.

* *Note:* Safety should be addressed *throughout* the manual.

SUMMARY

Be aware of the manual users as you plan your manual. Check back with your user profiles or characteristics and the environment of the manual's use as you write, revise, and update. The following steps will sharpen your awareness of the manual users and help to organize your manual.

- Think systematically about your manual users.
- Anticipate user questions.
- Use user questions to structure the manual: answers become sections.
- Employ user feedback or usability testing of manuals to revise, refine, and update your documents.

CHECKLIST: USER CHARACTERISTICS

Answer these user questions about the manuals you and your company produce. After you have answered the questions, consider whether you would want to change anything about the way your manual is put together.

Personal Characteristics of Users

- ☐ Do they use this machine or product almost every day or only once in a while?
- ☐ Are they likely to have used other products like it?
- ☐ Do they do their own routine maintenance?
- ☐ Do they do their own repairs? Should they?
- ☐ Do they understand technical language?
- ☐ Do they understand charts, circuit diagrams, mechanical drawings?

Conditions of Manual Use

- ☐ Will they use the manual only to learn how to set up, use, or operate the product?
- ☐ Will they refer to the manual or instructions often?
- ☐ Will they use the manual or instructions only if something breaks, fails, or appears to be abnormal?
- ☐ Will they read the entire manual or only a section here and there?
- ☐ Will they be able to look at the machine or product when they are reading the manual?
- ☐ Will the illumination be good?
- ☐ What are other physical conditions of use (wet, dry, oily, noisy, dusty, cramped, etc.)?

Information Wanted

- ☐ Basic instructions for use, operation, and adjustment?
- ☐ Routine maintenance procedures?
- ☐ Sophisticated service procedures?
- ☐ Specifications and parts lists?
- ☐ Troubleshooting procedures?
- ☐ Explanations of new technologies or product features?

REFERENCES

Coe, Marlana. *Human Factors for Technical Communicators*, John Wiley & Sons, New York, 1996.

Cooper, Alan. *The Inmates Are Running the Asylum: Why High-Tech Products Drive Us Crazy and How to Restore the Sanity,* SAMS, a division of Macmillan Computer Publishing, Indianapolis, 1999.

Laux, Lila. A Human Factors Approach to Developing and Evaluating Facilitators. Presentations 1997–1999 at "The Role of Warnings and Instructions" seminars, University of Wisconsin–Madison.

Norman, Donald A. *The Design of Everyday Things.* Doubleday/Currency, New York, 1990 (originally published as *The Psychology of Everyday Things,* Basic Books, New York, 1988).

Ranous, Charles A. Checklist developed at the University of Wisconsin–Madison in the 1970s.

Schriver, Karen A. *Dynamics in Document Design,* John Wiley & Sons, New York, 1997.

CHAPTER 3

Structuring the Manual

OVERVIEW

People tend to acquaint themselves with an unfamiliar product in much the same manner they find their way around a new city. They look at maps and search for landmarks and street signs. As you develop a manual, you can simulate the function of street signs, landmarks, and maps by careful organization and by writing strategies that help readers find their way around the manual and the product.

Few of us like to remain complete tourists in a place we visit often. We like to learn our way around and become familiar with major features so we can feel moderately comfortable traveling on our own. We feel much the same way about products and manuals. We use manuals, in fact, only to feel comfortable with the product. We want to use the product and not feel too foolish. A good manual helps us do both. And as with the map or guidebook, once we feel confident enough to cruise without carrying it everywhere, we put aside the manual, consulting it only now and then, when we're getting lost or desperate.

In this chapter we'll discuss what manuals need to provide, what you need to keep in mind and return to as you create a manual, and what specific strategies you can use to write and yet remain relatively sane in the process.

MANUALS ARE NOT NOVELS

Let's begin with the obvious, because too many manuals we've looked at don't: *manuals are not novels.*

True, manuals are composed of words on a page, just like a novel, but there the similarity begins and ends. Manuals are not read cover to cover, in a leisurely manner, for pleasure, or, as with mystery novels, with the vital information — Hortense was the killer all along — delayed until the end. Manuals are skimmed, read intently only sporadically and in parts, read with the reader in need and anxious and/or under stress, and, if vital information is not up front and immediate, perhaps not read at all. Manuals are not novels.

Manuals are more like guidebooks or cookbooks. They are, in fact, read at all only because we can't figure out the product by itself without them. In other words, we are needy and desperate. We are also usually in a hurry. After all, we really want to tape tonight's movie, not read a manual on how to set up the VCR. Reading a manual, then, is a means to an end. It is not the end itself. When you create a manual, you must plan for discontinuity, interruption, and stress. Like the writing process, the reading process of a manual is not linear — and not neat.

WHAT DO MANUALS OFFER THE USER?

Catch and Connect

Users come to manuals with questions. What do I need to do? What is this thing? How do these two fit? What goes where? If I do this, then what happens? What do I do next? How long is this going to take? What can I skip? Manuals need to "catch and connect" almost immediately: "catch the attention of busy readers," as Schriver explains, and "connect ... with the readers' knowledge, experience, beliefs, and values" (Schriver 166).

Pattern and Progress

Manuals also need to give users two basic comforts: a recognition of pattern and a sense of progress. Pattern comes when users see information that is clear, consistent, complete, coherent, and connected. In other words, information that has been focused and shaped for them. Progress comes when users sense movement and closure, when they believe there is a direction to their efforts and a definite end point. The manual brings them closer to effectively using the product. If users believe that, and trust you to get them there, they'll follow you anywhere.

HOW CAN YOU ORGANIZE AND WRITE MANUALS?

Create for the customer. It's as simple, and as complicated, as that. For each piece of information, verbal or visual, you place in the manual, pretend you are a customer of your own product and ask: So what? Why are you telling me this? What's in it for me? What do I have to *do*?

Remember that a manual is a guidebook or a cookbook. It's a tool to help the reader *do something else* — and something physical, at that: tour a city, bake a cake, use a chain saw to cut wood, or phone a friend. Readers of the manual are *users of the product*. They do not read for the act of reading itself. They read to help themselves do something with the product they can't do by themselves (or by asking someone else) without reading the manual. Users read because they need. Focus on this need to know.

A MANUAL IS NOT A LIBRARY

Focus on need-to-know information and leave out what's nice to know. Good organization of a manual means choosing what to leave out as well as what to put in. Manual writers with technical backgrounds, for example, are often tempted to include too much theoretical or technical material. The theory or experimental design that underlies a product will always be of interest to the manufacturer as well as to the technically sophisticated user. However, in manuals for the general public, less is more than enough. You should be guided by what the users need to know to use the product, not by what would be nice for them to think about.

Not only does excessive technical explanation take up page space and increase manual costs, it is also intimidating and distracting for beginning users of a product. How would you feel if you encountered any of the following samples of theoretical/technical material in a manual?

- A manual for a device that measures surface tension in which the first four pages describe the history of surface tension study, going all the way back to Archimedes
- A manual for an inexpensive stereo system that describes advances in electronics, printed circuits, and acoustics
- A manual for a sailboat that devotes over half its space to unresolved controversies on vector theory and wind motion

You must, of course, use your judgment to determine how much theory is appropriate for adequate understanding of your product. But remember that the average buyer of a microwave oven, for example, will probably not need an elaborate discussion of the electromagnetic spectrum and the esoteric physics of microwaves. (Some of this information might appear in a service manual, however. See Chapter 7.) The user only wants to heat a cup of coffee. Focus on the need to know.

STRATEGIES THAT WORK

At this point you probably need to know some concrete, physical things you can do to a text or on a page to create a manual that is genuinely helpful to the user. A manual that is easy for someone to use wears its skeleton on the outside. It always makes its structure visible to the reader.

Here are some strategies to help you make the skeleton obvious, the structure visible.

- Give cues.
- Repeat what's important.
- Build doors and ramps.
- Reduce text.
- Repeat what's important.

These strategies are seldom used in isolation. Rather, you'll probably combine techniques to ensure smooth flow and internal consistency. Again, remember that the point is to make the structure of the manual visible to the user.

Give Cues

Cues are landmarks, signposts, or (information) highway markers. Cues let the users know where they are and where they can go in a manual. Helpful cues you can build into your manual are as follows:

- Table of Contents and Index (more in Chapter 4)
- Headings — of sections, of subsections, of steps
- Overviews/introductions and summaries — of sections, of main clusters of paragraphs
- Transitions — repeated words and phrases to direct flow and provide connection or coherence from paragraph to paragraph or sentence to sentence
- Parallels, series, comparison–contrast
- Numbering of steps
- Running visuals — visuals that repeat to link a series of steps

Headings

General headings — labels that identify major sections — are the most common and immediately recognized cues. They provide landmarks for user understanding of major systems of a machine, major uses of a product, and/or major steps in a process. As suggested in Chapter 2, you can employ user questions to determine the major sections of a manual. Again, those major sections would often include the following:

- Scope
- Description
- Operation and Intended Use
- Special Features
- Limits of Operation
- Setting Up and Turning On
- Normal Operation
- Turning Off and Disposal
- Abnormal Operation (Troubleshooting)
- Preventive Maintenance
- Storage
- Safety (throughout the manual)

Most manuals will have some, if not all, of these sections, and you may choose to vary the number, order, and length of major sections to match the requirements of your product. The following are some examples of the general headings used to designate chapters or major sections of manuals for a variety of products.

Examples of General Headings

Surgical Aspirator
Scope and Purpose of Manual
Introduction to Model 923
Description and Function of Handpieces
Description and Function of Console
Unipaks and Setup

Herbicide
Precautions
Uses of Product
Mixing and Spraying
Fluid Test Compatibility
Application Information
Cultivation Information
Weed Control

Tractor
Safety
Controls and Instruments
Pre-start Checks
Operating the Tractor
Drawbar and PTO
Ballast
Transporting
Wheels, Tires, Treads
Lights and Signals
Fuels and Lubricants
Lubrication and Maintenance
Service
Storage
Troubleshooting
Lubrication Chart

Fabric Dye (home use)
Fabrics Suitable for Dye
Preparation of Fabric
Mixing and Preparing Dye
Applying the Dye
Fabric Care and Washing

Camping Tent
Capacity of Tent, Users
Parts of Tent
Assembly
Disassembly
Care of Tent
Storage

Table Saw
Safety
Unpacking and Cleaning
Connecting Saw to Power Source
Controls and Adjustments
Operation
Maintenance
Parts, Service Warranty

The examples show both chronology and spatial logic and, of course, reflect the complexity of the product. For instance, the herbicide manual is a small, pocket-size booklet of 45 pages, the tractor manual is 102 pages, and the tent manual is 8 pages. As the length of the manual increases, the general headings become more important as road signs for the user. The headings are also invaluable for you as you develop the outline for your manual.

Using Headings as Hierarchical Cues

Headings not only tell the reader the main idea of the section to follow, but also provide cues on the relative importance of various kinds of information. These cues are largely responses to print size, typeface, uppercase or lowercase letters, and position of the heading on the page. Individuals may disagree here and there on the meanings of these cues, but, in general, these responses hold true:

1. *Largest level of organization or most important information*
    - Larger type
    - Bolder print
    - All uppercase letters
    - Centered on the page
2. *Smaller level of organization*
    - Smaller type
    - Finer print
    - Mix of uppercase and lowercase letters, or all lowercase
    - Flush left or indented
3. *Special emphasis*
    - Italics or underline or bold
    - Boxes
    - Color

For example, a section of a manual laid out in the following way will be perceived by the reader to be arranged in a descending hierarchy of ideas.

Controls
On–Off
Pre-set
Throttle
Turning
- Right
- Left
- Circle

You can play with position on the page, upper- and lowercase, and bold, for example, in innovative ways — as long as you give the reader consistent messages throughout the manual. The cues you select train the reader to approach the material in certain ways. Define those ways for yourself and then clearly communicate them to the reader.

Choosing the Cues

Keep in mind that page design is an art. Desktop publishing has made manual production easier for many companies. It has not, however, necessarily increased or guaranteed aesthetic quality.

The increase in flexibility and choice provided by desktop publishing is both bane and blessing. You have many more ways to play with page design, but doing so *effectively* takes time: you need to develop particular skills to produce quality design. And be wary of that all-too-common design disease, "font futzing": just because you can see how the text looks in 27 different type styles and sizes doesn't mean you should. Professional graphic designers, printers, and publishers do have special skills you simply can't develop overnight, and the choices they, or you, make have effects on how the user sees the manual. (See Chapter 4 for more information on page design.)

Look at Figure 3.1 to see the power of print size, uppercase and lowercase, and boxes. The original also makes effective use of color. When manual users see a page like this, they perceive the following hierarchies:

1. General heading: STORAGE
   - The first major section of the manual
   - Separated from the text by full-page underscore line
   - "Storage" repeated at the bottom right of the page in italics on all subsequent pages that belong to this major section (not shown)
   - Very large outline type indicating further that "Storage" is a major section
2. Subheads: STORING THE TRACTOR, STORING BATTERIES, REMOVAL FROM STORAGE
   - Subhead type sized somewhat smaller than general heading, but larger than the type size for subsystems that fall under each subhead
   - All uppercase letters reinforcing hierarchical importance
3. Subsystems: Preparation, Storing
   - Combination of upper- and lowercase letters plus still smaller print size indicating lower hierarchical level
4. Boldface: Used for subheads and subsystems
   - Reinforces distinction between headings and the text proper
5. Boxes: Used for safety warnings
   - Boxes plus safety alert symbol in color (in the original) calling special attention to fire hazard

Subheads and Focusing Sentences as Cues

Sections or chapters can also use internal cues such as subheads, focusing sentences, and transitions to mark a reader's place and provide direction. Example 3.1 uses a general heading to indicate the section or topic and three subheads to identify the individual steps.

The first sentence of the section focuses attention and forecasts what follows. The sentence cues the reader how the subsequent paragraphs will be developed. The word "three" in sentence 1 and the list of three steps in sentence 2 are picked up and repeated, in the same order, in the subheads that follow. Numbering the steps, since you want them to be done in a particular order, would add another level of helpful cues.

5 Storage

STORING THE TRACTOR

Preparation

1. Change hydraulic oil.
2. Change transmission oil.
3. Change engine coolant.
4. Drain and flush gear oil from differentials and planetary gear housings. Fill with new oil.
5. Change engine oil and filters.
6. Start engine. While engine is warming up, operate the transmission, hydraulic system, steering and differentials to distribute the new lubricants to components. Warm engine to at least 70° C (160° F); it may be necessary to shield the radiator to achieve this temperature. Stop engine.
7. Clean tractor of all debris, dirt and accumulated grease.
8. Drive tractor to storage location.
9. Relieve tension on alternator, air conditioner compressor and fan belts.
10. Coat all exposed hydraulic cylinder shaft areas with grease or a rust preventive.

Storing

1. Using plastic bags or tape seal the following openings: air cleaner inlet, exhaust muffler, fuel tank breather and air conditioner air intake screens.
2. Touch up all scratched or chipped painted areas.
3. Block up tractor to remove weight from tires.
4. Cover tires if they will be exposed to heat and/or direct sunlight.
5. If tractor is to be stored outside, cover with a waterproof canvas or other protective material.

STORING BATTERIES

WARNING

AVOID SMOKING OR OPEN FLAMES IN OR NEAR BATTERY CHARGING AREA DURING OR FOR TWO HOURS FOLLOWING CHARGING.

1. Low maintenance batteries do not require charging before or during storage. Under normal conditions, storage life will be 12 months before recharging.
2. Check battery charge. If not 1.270 specific gravity, charge batteries. See ELECTRICAL, Section 3.
3. Remove batteries from tractor and store in a dry weatherproof area.

REMOVAL FROM STORAGE

1. Remove protective covering from tractor tires and seals from air cleaner inlet, exhaust muffler, fuel tank breather and air conditioner air intake screens.
2. Remove blocks. Lower tractor onto tires.
3. Correct any leaks.
4. Inflate tires to recommended pressure.
5. Install fully charged batteries. Tighten cable clamps at both ends of cables.
6. Tension alternator, air conditioner compressor and fan belts.

Figure 3.1 Effective use of general headings. This figure shows the power of print size, uppercase and lowercase letters, boxes. The original also makes effective use of color. (From *Service Manual, Series 2 Four-Wheel Drive Tractors,* Applicability: 1977 Production, Versatile Manufacturing Ltd., Winnipeg, Manitoba, Canada, 1977, 55. With permission.)

These cues are simple and consistent (if you promise the reader three, you need to deliver three) — and remarkably effective. They are the road signs that help the reader feel, "Okay, I'm still with you. I'm not getting lost."

Example 3.1 Paragraph cluster with subheads and core sentences.

OPERATION OF THE SPECTROPHOTOMETER

To operate the Ace Model X Spectrophotometer, three procedures must be followed. These include preliminary steps, placement of samples in the cell compartment, and measurement of samples.

Preliminary Steps

Before making any measurements, turn the sensitivity switch to standby, the shutter switch to SHTR, and the power and lamp switches to the ON position. Then allow the machine to warm up for 15 minutes before attempting any measurements....

Placement of Samples in Cell Compartment

Each machine has two rectangular-shaped test tubes called cells. One of these cells contains a reference sample. This sample ...

Measurement of the Samples

Slide the cell holder into position so that the reference cell is in the light path of the compartment. Select the desired wavelength ...

Other Focusing and Clustering Cues

Parallels — When procedures or mechanisms are closely related, parallel sentence strategies also serve to sharpen focus. Notice the difference in clarity in the following examples.

Version A: Nonparallel
Install front bolts with the threads down. On the rear bolts, make sure the threads face up.

Version B: Parallel
Install front bolts with threads down.
Install rear bolts with threads up.

Readers can understand instruction A, but they will grasp B more quickly.

If you are trying to make comparison or contrasts between steps or characteristics of a process or product, you can also use parallel structure to heighten the comparison.

Version A: Nonparallel comparison
More modern mechanisms have electronic controls, whereas they were formerly operated manually.

Version B: Parallel comparison
Modern mechanisms are electronically controlled, whereas formerly mechanisms were manually controlled.

Version A is comprehensible, but the comparison is not sharp. Version B is much sharper.

Series — If you are listing a series of parts or steps, keep the series grammatically consistent.

> **Version A: Inconsistent verb forms**
> Always wait for the tractor to come to a complete stop. After lowering the equipment to the ground, make sure the transmission is shifted to the N position; the park brake should be set to prevent the tractor from rolling. Then remove the key.

> **Version B: Consistent verb forms**
> Always wait for the tractor to come to a complete stop, then lower equipment to the ground, shift the transmission to N position, set the park brake so the tractor will not roll, and remove the key.

Readers will understand version A, but they will have to slog through shifts from active to passive voice and unnecessary glitches in verb tense sequence ("wait," "after lowering," "is shifted," "should be set," "remove"). In version B, each element of the series begins with an imperative verb ("wait," "lower," "shift," "set," "remove"). The series of commands is consistent, predictable, comfortable — in short, a clear and reassuring map of what to do.

Repeat What's Important

You emphasize what's important in a manual by giving it higher-level headings, by putting it first, by talking more about it, or by using typographical cues such as boldface, centering, or white space. You can also emphasize what's important, or what the user might refer to again and again, by repeating it. For example, you can repeat critical information in a segment summary, in a checklist at the end of a section, or as a full section in the appendix.

Repetition is particularly important in the case of safety warnings. If you are warning of danger and death, once is not enough. A full-page list of all the warnings, placed in the front of the manual (and never referred to again), is certainly not enough — and probably not effective, judging by how many people say they skip over that page entirely. Safety warnings should occur throughout the manual and be repeated, depending on the severity of the danger, on the product or equipment. Chapter 6 discusses safety warnings in greater detail, and with examples.

Build Doors and Ramps

If you think of manuals as houses, imagine doors. If you think of manuals as highways, imagine lots of on-ramps and exits. You want to provide access, facilitate flow, create movement through the information in your manual. You do this by providing a floor plan or map — the Table of Contents and the Index — and by organizing sections as stand-alone yet sequential modules, where users can drop in and out of sections as needed.

Make Your Writing Modular

You can carry this design further by writing short, focused, unified, and coherent paragraphs, and short, focused, verb-first (active-voice) sentences. In other words, design everything, from sections/chapters to sentences, in chunks. Modular, short, and focused means easy entry, easy exit, for the reader. Since the reader is often using the manual and working with the product at the same time, finding his or her place and moving from place to place easily are very important.

Sequence Material Clearly

Another way to help the reader move through the manual is by careful sequencing of information. We'll discuss three major kinds of sequencing: general to specific, spatial, and chronological.

General to Specific — We have said that users respond to unfamiliar products and their manuals by searching for landmarks. Another way to think about this is to consider how you respond to meeting a stranger. You begin with a general impression of a person strange to you and only later begin to notice and understand details of dress, manner, or speech. The principle is a simple one: it is psychologically more natural for most people to move from general to specific. Knowing this, you should try to introduce your user to the product by arranging the material of the manual from general to particular.

One way to assure this progression from general to specific is to think of your manual in terms of overview first, then details.

Suppose, for example, that you want to provide instructions for a stereo system. The first section of the manual should contain an overview (a labeled, full-shot photograph or drawing, a verbal description, or both) of the entire system. This overview gives the reader a sense of how the parts relate to each other. Then by means of details (subsystems), you may begin to provide more-detailed information for each of the parts.

This same general-to-specific strategy also holds true for descriptions of processes or procedures. Below are two outlines that will give you a rough pattern for describing a mechanism and a process.

Description of a mechanism

1. Introduction
   a. Definition
   b. Purpose
   c. General description
   d. Division of a device into its components
2. Principle or theory of operation
   a. Divisions — what the part is, its purpose, and its appearance
   b. Divisions into subparts
      i. Purpose
      ii. Appearance — often through visuals
      iii. Details — shape, size, relationship to other parts, connections, material

3. Operation of the system
   a. Ways in which each division achieves its purpose
   b. Causes and effects of the device in operation

Description of a process
1. Introduction
   a. General information regarding why, where, when, by whom, and in what way the process is performed or occurs
   b. List of the main steps
   c. List of the components involved
2. Description of the steps or analysis of the action
   a. First main step (or sequence of events)
      i. Definition
      ii. Special materials
   b. Division in substeps
3. Conclusion (summary statement about the purpose, operation, and evaluation of the whole process)

Spatial — For very large products, such as industrial machines, trucks, tractors, cranes, and even motorcycles, users say they approach the product with a preconceived spatial logic. For example, they may think of the product from front to back, from top to bottom, or from left to right when facing the product. Of course, variations of this basic perception will be required in specialized manuals or service instructions. In any case, visualize from the point of view of the product user. For example, the logical view for service manuals on exhaust systems would be to show and describe the system as seen from below by the mechanic as he or she works with the system overhead on a hoist.

Whatever the product or process, it helps to ask your person-on-the-street user, "How do you think of this? Do you stand in front or at the side? Do you think of this from front to back? Top to bottom?" Then arrange the manual to match the user's spatial perception. Do not, for example, begin with the rear axle assembly on a truck, or tell the user of a sewing machine how to embroider or buttonhole before you explain how to thread the machine or sew a straight seam.

Chronological — Most processes and procedures have an inherent chronology, i.e., the steps for doing something grow naturally out of the way the product or process works. For example, the user will usually want to know about setup or assembly before operating procedures, maintenance, or storage.

However, at the level of paragraph or subsection, exceptions to strict chronology are quite common. For example, suppose you are describing the operation and use of a home whirlpool. The dangers posed by high water temperatures to the elderly and to people with heart conditions or high blood pressure need to be mentioned, both in the manual and on the product, *before* the owner uses the whirlpool.

Likewise, you would not instruct a user how to remove the cover of a pressure cooker without first showing how to make sure that all the steam has escaped from the cooler. In short, if any step or procedure can, in its execution, cause damage to the product or injury to the user, be sure to explain this before the step is listed.

Anticipate trouble spots in procedures, even if chronology is interrupted. *Warn of troubles or dangers before it is too late for the user to do something about them.*

Example 3.2 shows two sets of instructions, one that maintains strict chronology, to the possible detriment of the user, and another that violates strict chronology but gives a timely warning.

Example 3.2 Maintaining proper chronology.

Finishing Your Home-Built Furniture

Advice too late

1. Prepare wood surface by using fine sandpaper or steel wool.
2. Mix oil-based stain of your choice.
3. Apply stain thinly with brush or cloth.
4. Allow to stand 5 minutes.
5. Remove stain by wiping.
6. Allow stained surface to dry for 24 hours.
7. Apply final finish of urethane or varnish.

Note: Some soft woods absorb stain quickly. If you are unsure of your wood, apply stain to test area first.

Advice on time

Many oil-based stains work best on hardwoods like cherry, walnut, or oak. If your furniture is made of soft wood, such as pine, test a sample area first to check absorbency of stain. If stain soaks or smears, prime surface with thinned shellac before finishing.

(Thinned shellac: 2 parts solvent/1 part shellac)

1. Prepare wood surface ...
2. Test sample area for stain absorption.
3. Apply thinned shellac primer if necessary and allow to dry thoroughly.
4. Apply stain thinly ...

Reduce Text

This may seem like a funny thing to say to writers, but it's true. Wherever possible, reduce the bulk of verbal text in your manual. Focus the reader's attention on need-to-know information only, and make it as easy as possible for them to move through — and out of — your document. You can reduce the text in a number of ways:

Go visual (see Chapter 5 for details)

- Use visuals to complement and/or replace text.
- Use charts, tables, graphs, especially for process/flow and troubleshooting.

Go for the verb

- Go short in paragraphs and sentences.
- Use active-voice/verb-first: "verb the text."

Go vertical

- Make lists.
- Make more lists.

The list strategy, in particular, is one of your most valuable tools. Many manuals rely too heavily on the linear mode, stringing out a series of instructions horizontally in long, complicated sentences and paragraphs. The horizontal arrangement makes instructions more difficult to follow. Far clearer to go vertical.

For example, if a novice photographer is developing film for the first time in a home darkroom, he or she will be following a series of steps very closely. Seeing those steps numbered and arranged vertically helps users to keep their place. The user tracks the process by thinking, "Step 1 finished. Now for step 2."

The list strategy makes use of a powerful communication tool, the vertical. To understand the power of the vertical, add the following set: 456 + 1678 + 45 + 789 + 9. Keep track of how long it takes you to do this.

Now add:

```
 357
4789
  23
 540
   8
____
```

The number of units, tens, hundreds, and thousands is the same in both sets, but most people are far quicker with the vertical. The list is the verbal equivalent of the numerical set. In Example 3.3 the words *before* and *when* serve as predictable organizers, like units and tens, and the reader's eye picks up only the key words as it sweeps down the passage. Note also that the white space around each element makes the list even clearer.

Example 3.3 Linear mode vs. list strategy.

The following paragraph is written in *linear* mode:

The system must be vented under the following circumstances: Before starting an engine that has not been operated for an extended period of time. When the fuel filters have been replaced. When an engine, in operation, runs out of fuel. When any connections between the injection pump and fuel tank have been loosened or broken for any reason.

Rewritten, its focus is sharpened by the *list:*

The system must be vented under the following circumstances:

- Before starting an engine that has not been operated for an extended period of time
- When fuel filters have been replaced
- When an engine, in operation, runs out of fuel
- When any connections between injection pump and fuel tank have been loosened or broken

Repeat What's Important — Again

If the information is vital for safe and effective use, repeat it. Since the users are jumping into and out of the manual as they work with the product, and skipping back and forth through the sections, you need to remind them of vital information. You can't count on everyone seeing the one page on which you explained how to do three subsequent operations. Give an overview, and then give the necessary information again at exactly the point it's needed. Repeating need-to-know information is helpful to users because it focuses their attention and saves them time looking back through previous sections for what they need now.

Combining Strategies

The strategies described in preceding sections seldom occur in isolation. Rather, writers tend to combine the techniques to ensure smooth flow and internally consistent logic. Look at Figures 3.2 through 3.6 to see how writers have combined a number of strategies.

REVIEW OF EFFECTIVE WRITING STRATEGIES

Using Example 3.4, take a couple minutes to review your understanding of the writing strategies we've discussed in this chapter.

Example 3.4 Review of effective writing strategies: comparison of two sets of instructions.

Read these two sets of instructions for using a portable tape recorder/radio. How would you answer these questions?

1. Which set of instructions is easier to follow?
2. What writing strategies make that set easier?

Passage A

Before playing tapes, make sure the radio switch is turned off.

1. To open the cassette door, press the button that stops the player and ejects the tape.
2. Insert a cassette into position so that the tape side is facing upward and the tape itself is on the right. Then press the cassette into place securely.
3. Push the cassette door closed and press the play button.
4. If you want to adjust the sound, adjust the controls for volume level and tone.

Passage B

1. Set RADIO switch to OFF.
2. Press STOP/EJECT button to open the Cassette Door.
3. Insert cassette into position as shown – open tape side up and full reel toward the right. Press the back of the cassette all the way in.
4. Close the Cassette Door and press PLAY.
5. Adjust VOLUME and TONE for desired sound.

UNPACKING AND CLEANING

Carefully unpack the saw, stand and all loose items from the carton. Remove the protective coating from the saw table surface. This coating may be removed with a soft cloth moistened with kerosene (do not use acetone, gasoline or lacquer thinner for this purpose). After cleaning, cover the table surface with a good quality paste wax.

ASSEMBLY INSTRUCTIONS

WARNING: FOR YOUR OWN SAFETY, DO NOT CONNECT THE SAW TO THE POWER SOURCE UNTIL THE SAW IS COMPLETELY ASSEMBLED AND YOU HAVE READ AND UNDERSTOOD THE ENTIRE OWNERS MANUAL.

ASSEMBLING STAND

1. Assemble the two top side braces (A) Fig. 4, which are 16-1/2" long, and the two top front and rear braces (B), which are 19" long, to the four legs (C) using the sixteen 5/8" long carriage bolts, flat washers and hex nuts supplied. **NOTE:** The top lips of the two top side braces (A) must fit on top of the top lips of the front and rear braces (B). The side braces (A) have holes on top for mounting the saw to the stand. Only tighten hex nuts finger-tight at this time.

2. Assemble the two bottom side braces (D) Fig. 4, which are 20" long, and the two front and rear braces (E), which are 22-1/2" long, to the four legs (C) using the sixteen 5/8" long carriage bolts, flat washers and hex nuts supplied. Only tighten hex nuts finger-tight at this time.

3. Assemble the four rubber feet (F) Fig. 4, to the bottom of each leg (C) as shown.

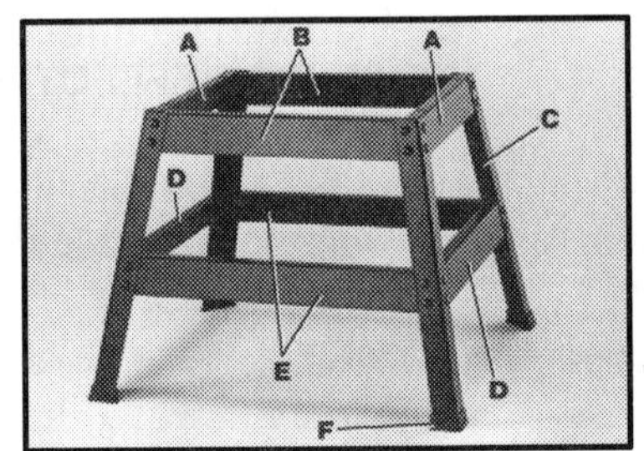

Fig. 4

ASSEMBLING SAW TO STAND

1. Position the saw (B) on the stand as shown in Fig. 5, lining up the four holes on the bottom of sides of the saw cabinet with the four holes in the two top side braces, one of which is shown at (A).

2. Fasten the saw to the stand using the four 5/8" long hex head screws, flat washers and hex nuts supplied.

3. After saw is assembled to stand, firmly tighten all stand mounting hardware.

Fig. 5

Figure 3.2 Combined writing strategies: general to particular, paragraph clusters, chronology. The sequence of pages follows the natural order of unpacking, cleaning, and assembly. Paragraph clusters are placed close to relevant visuals. Instructions are present-tense, active verbs. (From *10" Table Saw (Model 34-670) Instruction Manual, Part No. 1340213,* Delta International Machinery Corporation, Pittsburgh, PA, 1995, 5–7. With permission.)

ASSEMBLING BLADE RAISING AND TILTING HANDWHEELS

1. Assemble the blade raising handwheel (A) Fig. 6, to the blade raising screw (B) making sure the slots (C) in the hub of the handwheel are engaged with the roll pins (D) on the raising screw shaft.

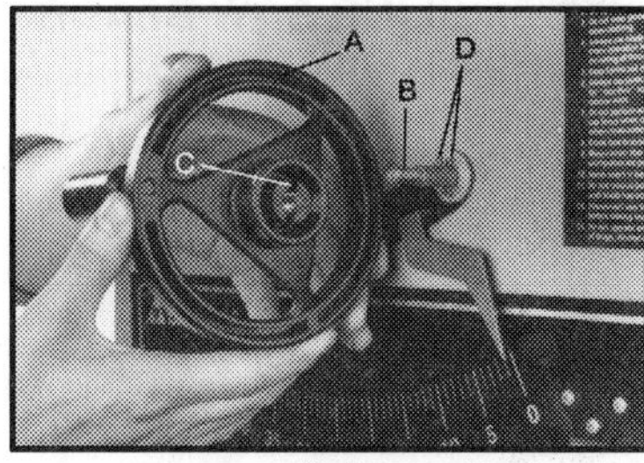

Fig. 6

2. Screw lock knob (E) Fig. 7, on end of raising screw shaft.

3. Assemble tilting screw handwheel (F) and lock knob (G) Fig. 7, to the blade tilting screw shaft in the same manner, as shown in Fig. 7.

Fig. 7

ASSEMBLING EXTENSION WINGS

1. Assemble extension wing (A) Fig. 8, to the saw table using the three screws and washers (B). With a straight edge (C) Fig. 9, make sure the extension wing is level with the saw table before tightening the three screws (B) Fig. 8.

2. Assemble the other extension wing to the opposite end of the table in the same manner.

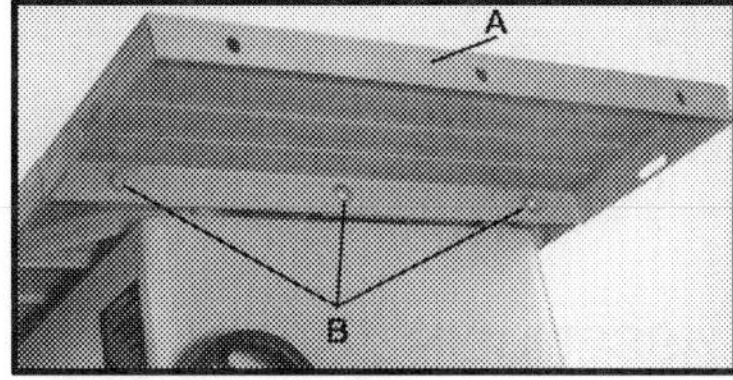

Fig. 8

Fig. 9

Figure 3.2 (continued).

ASSEMBLING SAW BLADE

1. Make certain the saw is disconnected from the power source.

2. Remove the table insert (A) Fig. 10.

3. Raise the saw blade arbor (B) Fig. 10, to its maximum height by turning the blade raising handwheel counter-clockwise and remove the arbor nut (E) and flange (D) from the saw arbor.

4. Assemble the saw blade (C) to the saw arbor making sure the teeth of the blade point down at the front of the table, as shown in Fig. 10, and assemble the flange (D) and arbor nut (E) to the saw arbor and tighten arbor nut (E) as far as possible by hand, being sure that the saw blade is against the inner blade flange.

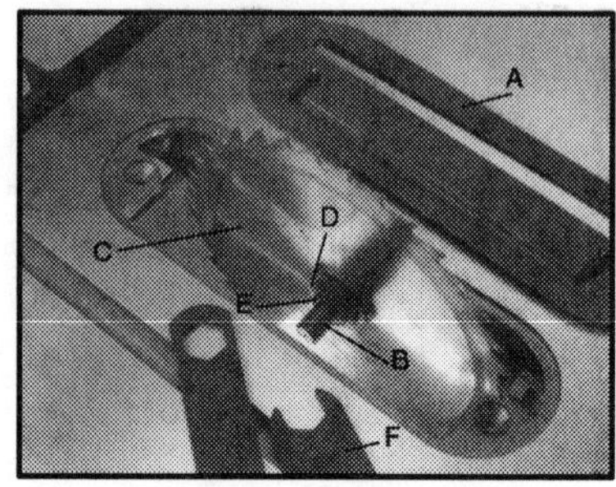

Fig. 10

5. Using the open end wrench (F) Fig. 10 and Fig. 11, supplied, place the wrench (F) on the flats on the saw arbor to keep the arbor from turning and tighten arbor nut (E) using the remaining wrench (G) Fig. 11, by turning the nut counterclockwise.

6. Replace table insert (A) Fig. 11, making certain that it is flush with table surface.

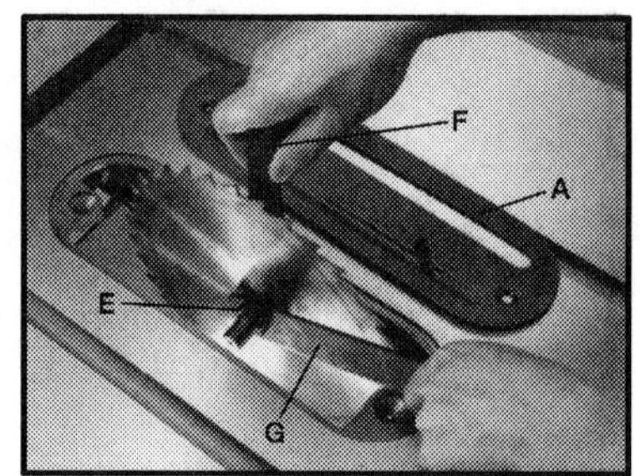

Fig. 11

ASSEMBLING GUIDE RAILS

1. The guide rail (A) Fig. 12, with graduations is to be assembled to the front of the saw table with the graduations up.

2. Insert the two special screws (B) Fig. 12, through the two holes (C) in the guide rail and place spacers (D) between the guide rail (A) and saw table. Thread the two special screws (B) into the tapped holes in the saw table. Do not completely tighten the two screws (B) at this time.

3. Insert special screw (E) Fig. 12, through hole (F) in guide rail and place spacer (G) between guide rail and extension wing. Fasten with flat washer, lock washer and nut (H). Tighten three screws (B) and (E) to fasten guide rail to table and extension wing.

4. Assemble the remaining guide rail to the rear of the table in the same manner.

Fig. 12

Figure 3.2 (continued).

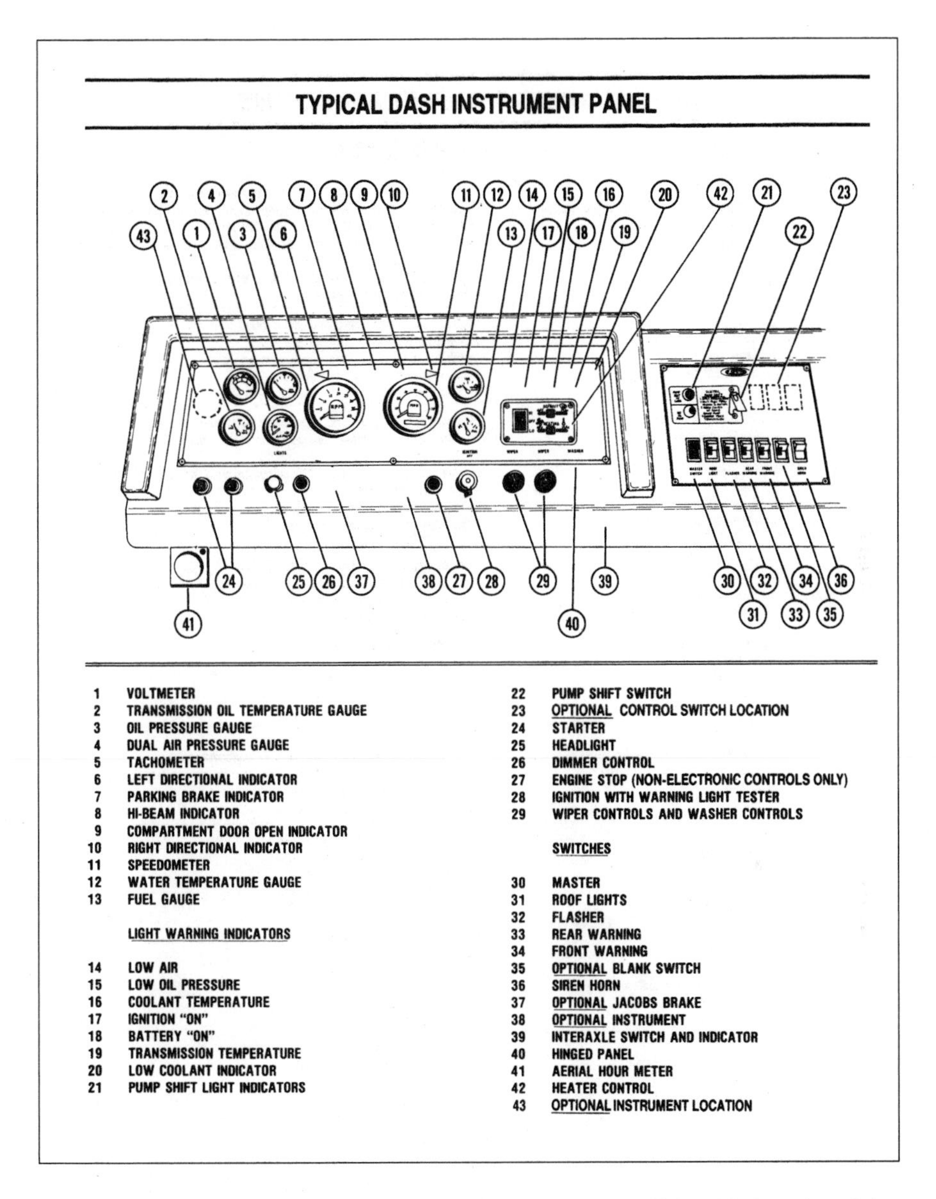

Figure 3.3. Combined strategies: clusters, boldface, glossary list. This "older" version of the manual uses a vertical format. Reader will look at diagram, then at callout, and then at explanation. Segmenting of long lists into groups of five makes them less forbidding to read and keep track of. (From *Pierce Chassis Operators Manual*, Pierce Manufacturing, Inc., Appleton, WI, 1990, 9–10. With permission.)

TYPICAL DASH INSTRUMENTS AND CONTROLS

NO.	IDENTIFICATION	NORMAL USE OR READING
1	Voltmeter	Indicates battery condition and rate of charge or discharge.
2	Main Transmission Oil Temperature Gauge	Registers main transmission oil temperature. Normal operating range is 160°-220°F (71°-104°C). Maximum is 250°F (121°C).
3	Oil Pressure Gauge	Indicates engine oil pressure in PSI. Stop engine immediately if low or no pressure is indicated. (5-70 PSI)
4	Dual Air Pressure Gauge	Air pressure should be from 90-120 PSI while operating.
5	Tachometer	Indicates engine speed (RPM).
6	Left Turn Signal Indicator	Flashes green when left turn signal is ON.
7	Parking Brake Indicator	Illuminates red when parking brake is set.
8	Hi-Beam Indicator Light	Illuminates blue when headlights are on high beam.
9	Compartment Door Open Indicator	Illuminates when door(s) is(are) open.
10	Right Turn Signal Indicator	Flashes green when right turn signal is ON.
11	Speedometer/Odometer	Indicates vehicle speed and records total accumulated mileage.
12	Water Temperature Gauge	Indicates cooling system temperature. (170°-195°F, 205°F; 77°-88°C, 96°C).
13	Fuel Gauge	Indicates level of fuel in tank. Fill fuel tank at the end of each day's operation to prevent condensation.
14	Low Air Indicator	Illuminates when air pressure is low.
15	Low Oil Pressure Indicator	Illuminates when oil pressure is low.
16	Coolant Temperature Indicator	Illuminates when cooling temperature is high.
17	Ignition Switch Indicator	Illuminates green when ignition switch is ON.
18	Battery ON Indicator	Illuminates green when battery switch is ON.
19	Transmission Temperature Indicator	Illuminates when oil temperature is high.
20	Low Coolant Indicator	Indicates loss of engine coolant.
21	Pump Shift Indicator Lights	Green: OK to pump. Red-flashing: Pump not in gear.
22	Pump Shift Switch	To shift pump in or out of gear.
23	Optional Switch Location	May be used for optional equipment.
24	Start Buttons	Push black buttons to start engine.
25	Headlight Switch	Pull ON push OFF for headlight control.
26	Dimmer Control	Turn to vary instrument light intensity.
27	Engine Stop (Non-Electronic Controls Only)	Push red button to stop engine (Non-electronic controls only).
28	Ignition Switch/Warning Light Tester	ON/OFF switch for engine electrical power. Turn switch completely clockwise.
29	Wiper Controls	Wiper—turn knob clockwise for ON and push to wash.
30	Master Switch	ON/OFF rocker type for electrical power.
31	Roof Light Switch	ON/OFF rocker type—push top for ON.
32	Flasher Switch	ON/OFF rocker type—push top for ON.
33	Rear Warning Switch	ON/OFF rocker type—push top for ON.
34	Front Warning Switch	ON/OFF rocker type—push top for ON.
35	Optional Blank Switch	ON/OFF rocker type—push top for ON.
36	Siren Horn	ON/ON rocker type—push top for siren; push bottom for horn.
37	Optional Jacobs Brake	Control location.
38	Optional Instrument	May be used for optional instrument.
39	Interaxle Switch and Indicator	Control location.
40	Hinged Gauge Panel	To access instrument panel for service.
41	Hour Meter (Aerial)	Records aerial hours of operation.
42	Heater Controls	Heater/Defroster temperature and fan control.
43	Optional Instrument Location	May be used for optional instrument

Figure 3.3 (continued).

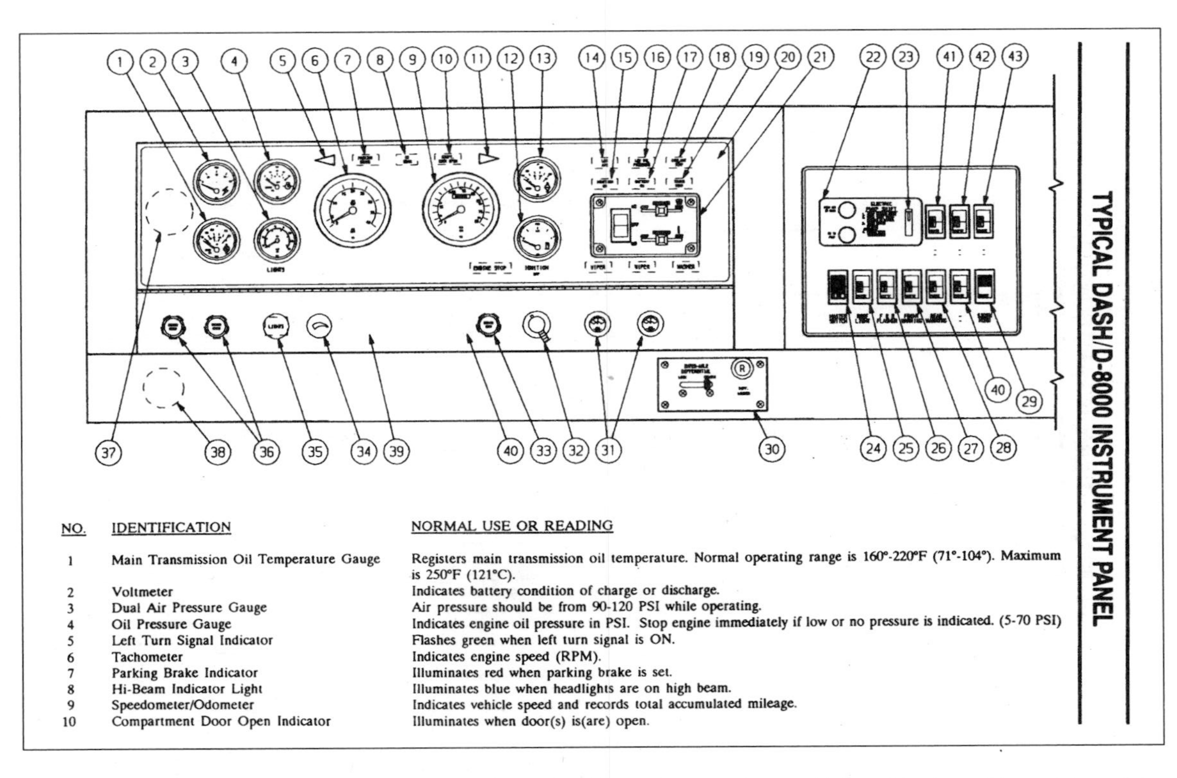

NO.	IDENTIFICATION	NORMAL USE OR READING
1	Main Transmission Oil Temperature Gauge	Registers main transmission oil temperature. Normal operating range is 160°-220°F (71°-104°). Maximum is 250°F (121°C).
2	Voltmeter	Indicates battery condition of charge or discharge.
3	Dual Air Pressure Gauge	Air pressure should be from 90-120 PSI while operating.
4	Oil Pressure Gauge	Indicates engine oil pressure in PSI. Stop engine immediately if low or no pressure is indicated. (5-70 PSI)
5	Left Turn Signal Indicator	Flashes green when left turn signal is ON.
6	Tachometer	Indicates engine speed (RPM).
7	Parking Brake Indicator	Illuminates red when parking brake is set.
8	Hi-Beam Indicator Light	Illuminates blue when headlights are on high beam.
9	Speedometer/Odometer	Indicates vehicle speed and records total accumulated mileage.
10	Compartment Door Open Indicator	Illuminates when door(s) is(are) open.

Figure 3.4 Revised version of pages shown in Figure 3.3. Uses clusters, glossary list, and horizontal format. In revising, the writer reduced the cross-reference job for the reader. Reader will now look at diagram and callout — two steps, rather than three. Horizontal format improves sequencing and alignment of numbers. (From *Pierce Chassis Operators Manual*, Pierce Manufacturing, Inc., Appleton, WI, 1990, 9–10. With permission.)

TYPICAL DASH/D-8000 INSTRUMENTS AND CONTROLS

11	Right Turn Signal Indicator	Flashes green when right turn signal is ON.
12	Fuel Gauge	Indicates level of fuel in tank. Fill fuel tank at the end of each day's operation to prevent condensation.
13	Water Temperature Gauge	Indicates cooling system temperature. (170°-195°F, 205°F; 77°-88°C, 96°C).
14	Low Air Indicator	Illuminates when air pressure is low.
15	Ignition Switch Indicator	Illuminates green when ignition switch is ON.
16	Battery ON Indicator	Illuminates green when battery switch is ON.
17	Low Oil Pressure Indicator	Illuminates when oil pressure is low.
18	Coolant Temperature Indicator	Illuminates when cooling temperature is high.
19	Transmission Temperature Indicator	Illuminates when oil temperature is high.
20	Low Coolant Indicator	Indicates loss of engine coolant.
21	Heater Controls	Heater/Defroster temperature and fan control.
22	Pump Shift Indicator Lights	Green: OK to pump. Red-flashing: Pump not in gear.
23	Pump Shift Switch	To shift pump in or out of gear.
24	Master Switch	ON/OFF rocker type for electrical power.
25	Roof Light Switch	ON/OFF rocker type-push top for ON.
26	Flasher Switch	ON/OFF rocker type-push top for ON.
27	Front Warning Switch	ON/OFF rocker type-push top for ON.
28	Rear Warning Switch	ON/OFF rocker type-push top for ON.
29	Siren Horn	ON/ON rocker type-push top for siren; push top for siren; push for horn.
30	Interaxle Switch and Indicator	Control location.
31	Wiper Controls	Wiper-turn knob clockwise for ON and push to wash.
32	Ignition Switch/Warning Light Tester	ON/OFF switch for engine electrical power. Turn switch completely clockwise.
33	Engine Stop (Non-Electronic Controls Only)	Push red button to stop engine (Non-electronic controls only).
34	Dimmer Control	Turn to vary instrument light intensity.
35	Headlight Switch	Pull ON push OFF for headlight control.
36	Start Buttons	Push black buttons to start engine.
37	Optional Instrument	
38	Optional Instrument	
39	Optional Jacobs Brake	Control location.
40	Optional Switch Location	
41	Optional Switch Location	
42	Optional Switch Location	
43	Optional Switch Location	

NOTE: Gauge Panel is hinged to access instrument panel for service.

Figure 3.4 (continued).

NM650

COMPONENT DESCRIPTION

Control Box: Contains the electrical controls that operate the machine.

High Pressure Cut Out Switch: A manual reset switch sensing the high side refrigeration pressure. It is set to shut the machine off, and illuminate the reset switch light if the discharge pressure should ever exceed 450 psig.

Compressor: The refrigerant vapor pump.

Reservoir: Float operated, it maintains the water level in the evaporator at a constant level, it also contains the water level sensor.

Water Level Sensor: Senses if there is water in the reservoir to make ice out of. Will shut the machine off it there is none.

Ice Discharge Chute: Directs the ice produced by the evaporator into the storage bin.

Ice Level Sensor: An electronic "eye", it senses the presence of ice in the bottom of the ice discharge chute. Operates to turn the ice machine on and off automatically as the level of ice in the bin changes.

Gear Motor: An oil filled, speed reduction gearbox, driving the auger.

Condenser: Air or water cooled, where the heat removed in ice making is discharged.

Expansion valve: The refrigerant metering device.

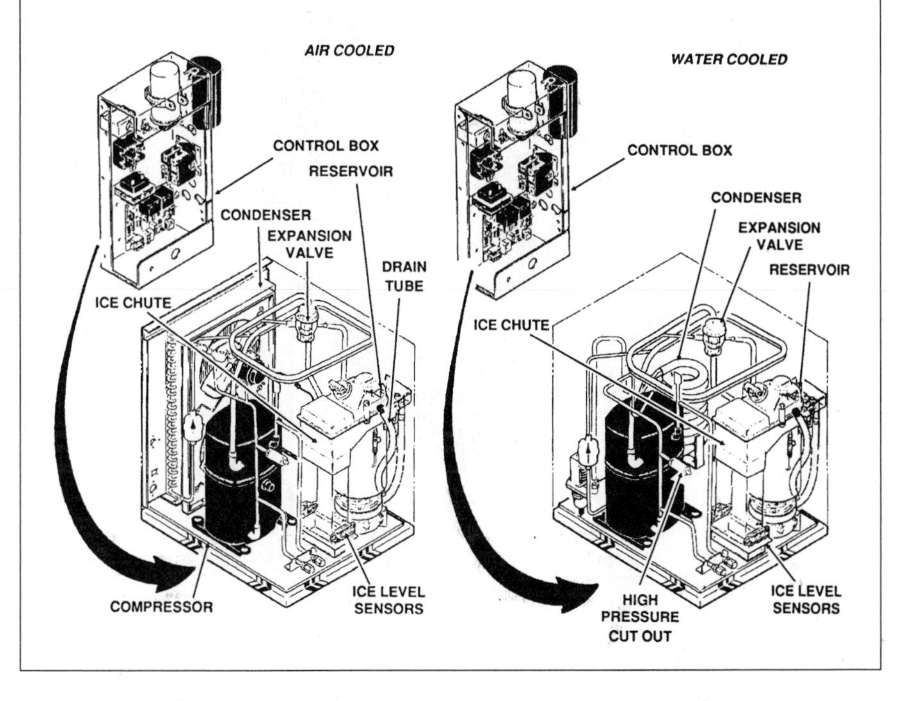

Figure 3.5 Combined strategies: general to particular, boldfaced list, glossary. Visuals and component lists on first page provide overview. Next page provides details for one component, the control box. (From *Service Manual for Modular Nugget Ice Maker Model MH 750*, Scotsman Ice Systems, Vernon Hills, IL, 10–11. With permission.)

NM650

COMPONENT DESCRIPTION: Control Box

Contactor: A definite purpose contactor connecting the compressor and the remote condenser fan motor to the power supply.

Circuit Board: Controlling the ice machine through sensors and relays. The sensors are for ice level and water level. The relays are for the gear motor (with a built in time delay to clear the evaporator of ice when the unit turns off) and for the compressor. The reset switch is mounted on the circuit board.

Transformer: Supplies low voltage to the circuit board.

Low Pressure Cut Out Switch: A manual reset control that shuts off the ice machine when the low side pressure drops below a preset point, 0-4 psig.

Potential Relay: The compressor start relay.

On/Off Switch: Manual control for the machine.

Reset Switch: Part of Circuit Board, manual reset. Lights up when unit shuts off from: ice discharge chute being overfilled (opening the microswitch at the top of the chute); low or high pressure switches opening.

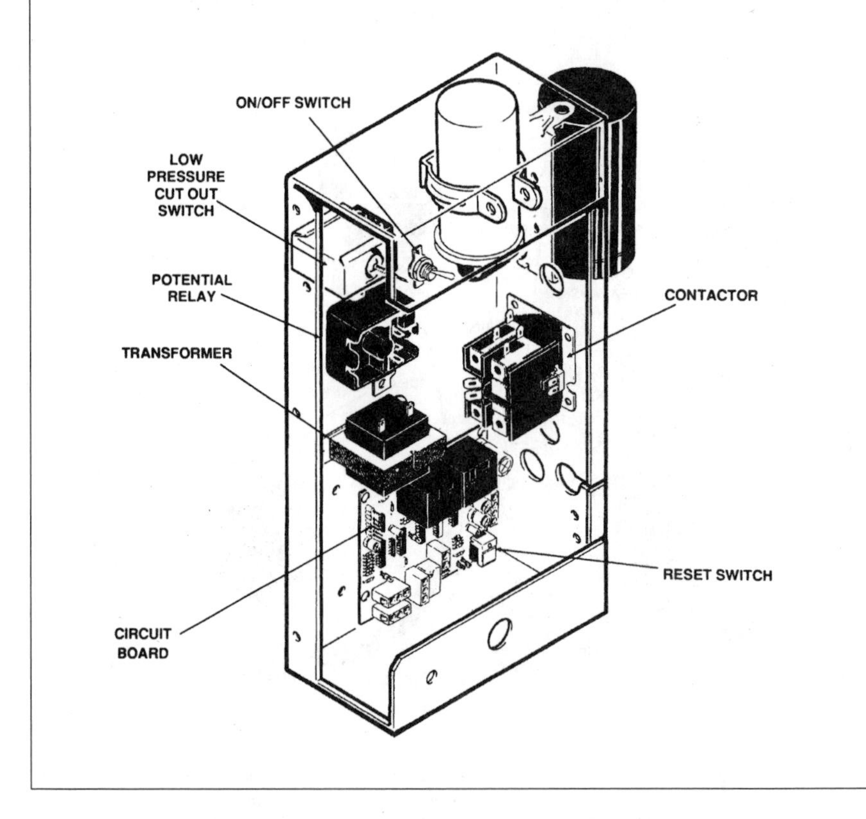

Figure 3.5 (continued).

Rear Panel Connectors

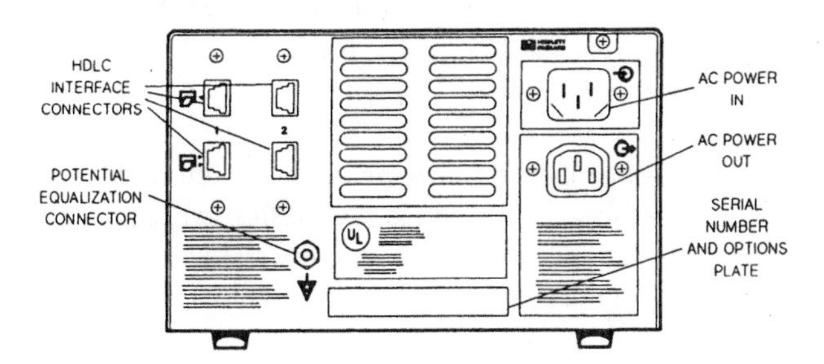

Figure 1-7. M1117A Rear Panel Connectors

HDLC Interface Connectors — The connectors in the upper left hand corner of the rear panel are used for communication between recorder and controller.

The "data in" connector is for signals coming from the controller and the "data out" connector is for re-transmitting controller data to another recorder.

Note The second set of HDLC connectors (the right two) are only present if option C01 has been ordered.

Potential Equalization Connector (Grounding Lug) — Safety Class 1 instruments are already included in the equipotential grounding system of the room by way of the protective grounding contacts in the power plug. However, for internal examinations of the heart, or the brain, the instruments must have a second connection to the equipotential grounding terminal on the instrument rear panel and the other end to one point of the equipotential grounding system. The equipotential grounding system assumes the safety function of the protective grounding conductor if ever there is a break in the protective grounding system.

Serial Number and Options Plate — The instrument's ten digit serial number is stamped here along with a code for any options which have been purchased.

AC Power In — One end of the power cable is plugged into this connector; the other end is plugged into a three-wire grounded wall outlet.

AC Power Out — The voltage at this connector is the same as the voltage that is applied at the power in connector. The AC power out connector may only be used to provide power to nearby instruments if certain requirements are met. In order to daisy chain power using this connector, the instrument receiving

Figure 3.6 Combined strategies. Products with keyboards, control panels, and graphics use glossary format for parts identification. (From *Model M1117A Multichannel Thermal Array Recorder*, Hewlett-Packard, 1989, 1–9. With permission.)

SUMMARY

Intelligent use of organization and writing strategies:

- Makes the structure of the manual jump off the page
- Provides important landmarks for the reader
- Reduces the bulk of the verbal text

First and last, always remember to *write for the user.* Tell users what they need to know; omit material that's merely nice to know. Design everything with this thought in mind: the user is almost always reading your manual while doing something else (like working with the product).

CHECKLIST

The following is a list of questions to ask yourself about the writing strategies for your manual.

- ☐ Have I identified what's important — from the user's point of view?
- ☐ Have I focused on need-to-know information?
- ☐ Have I repeated important information?
- ☐ Have I given cues, such as a table of contents, headings, overviews and summaries, and transitions, to help readers find their way through the manual?
- ☐ Have I made it easy for the user to skip around in the manual?
- ☐ Have I presented sections in easy-to-use modules?
- ☐ Have I ordered the sections in a logical sequence?
- ☐ Have I reduced the amount of text — by going visual, vertical, and for the verb — wherever possible?
- ☐ Have I used many graphics?
- ☐ Have I warned of safety issues before the hazardous step?
- ☐ Have I answered the user's questions?

REFERENCE

Schriver, Karen A. *Dynamics in Document Design,* John Wiley & Sons, New York, 1997.

CHAPTER 4

Designing the Manual

OVERVIEW

Effective manuals are not simply collections of good writing and clear illustrations. For the manual to be *used*, rather than left on the shelf, it must invite the reader to pick it up. Once the reader has opened the manual and started to read, the design must make it easy for the reader to find his or her way to the information needed. While good writing and clear illustrations are a necessity, much of what makes a manual attractive and easy to use comes from design decisions made before a word has been written or a line has been drawn. These decisions — about such "surface" elements as column width, type size, heading structure, as well as physical elements like the paper used and the kind of binding — may determine whether anyone reads the manual at all. The best writing and the finest illustrations are of no use if no one chooses to open the manual.

This chapter discusses how to design manuals that will be both useful and used. Creating publications that will compete successfully with all the other demands for a reader's attention is no easy task, but as products become more complex and users more diverse, effective documentation becomes crucial. The good news is that computer technology has made it possible for a small publications department (or even a single writer) to produce manuals of as high a quality as those from the largest corporation. This chapter explores these initial design decisions that must be made and offers some guidelines to help make sure that your manual makes it out of the shrink-wrap and off the shelf.

WRITERS AS DESIGNERS

When we wrote the first edition of this book back in 1984, we titled it *Writing and Designing Manuals* because at that time those were two separate functions performed by different people. Back then the norm was that writers wrote text, illustrators drew art, and designers — sometimes called "paste-up artists" — put the two together into page layouts. Indeed, the designers often literally "pasted up" page layouts, cutting "camera-ready" printed text and art with a razor blade and fitting it together on a "mechanical," which then was photographically reproduced for printing. Photographs, or "continuous-tone" art had to be photographed through a screen that converted the image to black dots of various sizes and spacings. The viewer's eye then would merge the dots and see them as shades of gray, the same way our eyes blend dots of color in an Impressionist painting. A published manual represented the combined efforts of a crew of specialists.

The computer has changed all that. In the short space of 15 years, an entire industry has been transformed, and terms like *halftone screen* and *Linotype operator* now seem as quaint as buggy whips. Now writers can choose a font with the click of a mouse button, artists can resize or reorient images electronically, photographs can be scanned, digitized, and stored as electronic files. The writer, as likely as not, will be the one to put it all together in a page design — electronically, on the computer of course. What used to be a team effort of a half a dozen or more specialists may now be a collaboration between two: a writer and an artist. (In some small operations, writer and artist may even be one and the same.) This change has provided both opportunity and challenge.

When the same person who writes the chapter also chooses the typeface and lays out the pages, that person has the opportunity to ensure that the arrangement of the text and art on the page reflects the meaning. The writer understands the organizational framework of the text and knows where the illustrations need to be placed for the reader to make best use of them. Good page design gives the reader subliminal clues to the meaning carried in the words and pictures: the writer, who knows the meaning backward and forward, is an ideal choice to arrange those subliminal clues.

The disadvantage, of course, is that the writer may not know much about design. One of the downsides of the computer revolution has been to give people with an untrained eye access to dozens of choices for typeface and whole libraries of clip art. We have all seen hideous examples of the smorgasbord approach to layout: six different styles of type in four different sizes sharing the same page with decorative borders and cartoon clip art. Fortunately, most publishing programs help out a novice with templates and defaults to get started.

Choosing Publishing Software

A number of publishing programs are available, as are various illustration programs. These range from "desktop publishing" systems designed for the home user or workplace newsletter editor up to extremely complex custom applications costing thousands of dollars. Most technical publications departments are somewhere in-between, using programs like Adobe FrameMaker™.* These midlevel programs provide the beginning designer with considerable help in the form of templates and wizards, and also offer a good deal of flexibility for the advanced designer. Increasingly, these programs are offering HTML conversion, so that materials originally designed

* FrameMaker is a trademark of Adobe Systems, Inc.

for print can be automatically converted for publication on the Web. Detailed software reviews are beyond the scope of this book (and would no doubt be outdated by its publication date), but we can give some guidance on what to look for in publishing software. The following list of questions should help you narrow the choices.

1. *What sort of computer systems does your company use now for other purposes?* If your company network is PC-based, using Windows NT, you may not want to choose a Macintosh publishing system, even if it would be better for your work. Ideally, you would want a system that could be integrated, so that, for example, drawings produced by the engineers on their CAD system could be imported and modified for insertion into a parts catalog.
2. *What kind of budget is available?* There is never as much money as you'd like, but how much can you realistically spend? Don't forget to include the cost of training for the people who will use the system. As a general rule, once you figure out what features you want, buy the best you can afford.
3. *What kinds of publications will you be creating?* Is your department responsible not only for producing manuals, but also for marketing materials such as color brochures? Be sure that the system you buy has the breadth you need.
4. *Does the system make it easy to revise or update manuals?* If you insert new text, does the system automatically reformat existing text to keep illustrations and text together, change figure numbering, appropriately revise cross-references, and so on? Or does all that need to be done manually?
5. *Does the system provide templates and defaults to help you get started in designing a manual?* At the same time, is it flexible enough to allow you to override the defaults and create your own look?
6. *How good is the technical support?* Is there built-in online help? Will the vendor provide technical assistance with an 800 number or the like? Will the vendor train you and your co-workers, or do you need to find independent instruction?
7. *Is the software compatible with the equipment used by your printing company?* Increasingly, documents are transferred from creator to publisher as electronic files, rather than as paper "paste-ups." Be sure that if this is the case in your company, your printer's equipment can read the files produced using your publishing software.
8. *Is the software user friendly?* There is always a learning curve to cope with when you are learning any new skill, but once you are past the initial phase, is the software easy or cumbersome to use? Think about the difference between early word-processing software and current models: early versions often required three or four keystrokes to make text appear in boldface; now that option is usually a button on a button bar. Both work — but one is a lot easier to use.

Design Basics

Much of the art of good page design must be learned by experience: both your own and other people's. You try something, and it works — or it doesn't. You notice a particularly effective layout in a document and spend a moment analyzing why it works. Gradually, you begin to get a feel for how various design elements work together and how to translate that information into placement of the text and illustrations ("copy" and "art") in your own manual. Nevertheless, we can identify some basic principles to keep in mind as you experiment.

The first of these, and the most important, is that *form follows function.* In other words, make your design choices on the basis of whether they make the manual work better. Remember that, in designing a manual, we are creating a document with a very specific purpose: to convey to the user the information needed to use the product properly and safely. Everything else comes second. If a design element does not contribute to that fundamental purpose, eliminate it — even if it is artistic, eye-catching, and innovative.

With that cardinal principle in mind, we can identify three basic qualities of good page design:

- A good design is simple.
- A good design is deliberate.
- A good design is consistent.

Simple

If the primary rule is that form follows function, then it stands to reason that the simplest form that will do the job is the best. Remember that the purpose of format and layout is to help the reader understand the information by signaling its structure. If the design becomes too complex, it will obscure the very thing it hopes to reveal. Too complicated a format makes it too hard for the reader to keep track of the various elements, and the reader will usually just give up trying. Most design innovations in products fall into two categories: those that improve performance and those that achieve the same performance, but with a simpler design. As long as your page design does its job, the simpler the better.

Deliberate

Product design engineers add parts to a product for specific purposes. Every nut and bolt, every spring, every lug, every piece of complexity is there to perform a particular function — if it is not needed, it is eliminated. The reason for this, of course, is that unneeded parts add to the cost of manufacturing and maintaining the product. In page design, "extra parts," that is, design elements that serve no function, or merely a decorative one, also add to the cost of the product, although in a more indirect way. Technical writing should be *transparent.* In other words, the writing (and, by extension, the page design) should convey the content as directly as possible to the reader, without calling attention to itself. In technical writing, the reader generally should not even be consciously aware of the language used or the design adopted. The focus should be on the information being conveyed, rather than on the form in which it is packaged.

If that is not the case, and the reader's attention is drawn to sentence structure or format, the reader is less likely to acquire the needed information to use the product properly and safely. Costs appear in the form of calls to the 800 number for technical support or, worst case, a product liability lawsuit. Is it a bit of a stretch to imagine that a flawed page design could be the ultimate cause of a product liability lawsuit? Yes and no. It is unlikely that page design alone would be at fault. However, it could easily be included as one more problem area in a faulty manual. Remember that a product may be deemed defective because of faulty instructions and warnings as well as because of defects in the product itself.

Consistent

Clearly, if the purpose of a format is to provide the reader with clues to how the material is organized, the format needs to be consistent or it won't work. If the writer uses 14-point bold headings in one place and 12-point italics in another to signal the same level of hierarchy in the organization, the reader will be confused. Since most manuals are team efforts, it is important to agree on formatting decisions early on, so that whatever is chosen can be applied by all the writers working on a single document. Likewise, if the same format is to be used in several different manuals, you must make sure that all the writers apply the format the same way.

Designing for Use: How People Actually Use Manuals

We said at the beginning that form should follow function. To design a good format for a manual, then, we need to know how that manual actually functions. In Chapter 2, "Analyzing the Manual Users," we addressed the question of who the users are: what sort of expertise they have, what kinds of environments they work in, and so on. The answers to those questions enable you to target your writing for the audience that will be reading it.

Choosing a page design or manual format requires a slightly different kind of audience analysis. This analysis focuses less on *who* the users are and more on *how* they use the manual. Interestingly, we can make some generalizations about manual use regardless of the user's level of expertise, familiarity with the product, and so on.

- They use multiple entry/exit points.
- They have short attention spans.
- They suffer from information overload.
- They use the manual in less than ideal environments.

Multiple Exit/Entry Points

Few users ever actually read a manual cover-to-cover. In fact, when we have polled the technical writers and artists attending our seminars on manual design, only about

10% of them routinely read the manual when they buy a new product — and these are people who produce manuals for a living. If they cannot be persuaded to read the whole manual, the average citizen certainly cannot. Instead, most people want to be able to get in, find the specific information they need at a particular time, and get back out.

Short Attention Spans

Most people's attention spans are getting shorter. We hear complaints about issues being reduced to a "10-second sound bite," yet that reflects a broader trend. Political ads that used to run 1 or 2 minutes (or more) are now almost universally only 30 seconds long. Newspapers like *USA Today*, which present short news summaries in a visually lively format, rather than in-depth stories in the traditional newspaper black-and-white, are national best-sellers. Television shows are interrupted ever more often by commercials that jump frenetically from shot to shot like music videos — a far cry from the "talking heads" of 20 years ago touting the benefits of this laundry detergent or that car wax.

Information Overload

Perhaps attention spans are shorter because the amount of information directed toward us is so overwhelming. For most of the time that humans have been present on Earth, change, particularly in our knowledge base, has come slowly. It has been measured in generations rather than years. The pace of change — and the accompanying need to learn new skills and comprehend new information — has accelerated wildly, and it shows no sign of slowing down. In the space of about 45 years, we have gone from describing the double-helix of DNA to inserting "designer genes" in plants to make them resistant to disease, not to mention cloning sheep. As recently as 40 years ago, a computer with less capacity than today's pocket calculator filled up an entire room. Today, using the Internet, we can communicate in real time (a term that did not exist 40 years ago) with someone halfway around the globe for the price of a local phone call. The technological advances have made it possible for just about anyone to have access to just about all the information on a topic that is available in the whole world. Is it any wonder we feel overwhelmed?

Less Than Ideal Conditions

As if all this weren't enough, people frequently need to use manuals under conditions that are not ideal for reading. If the user is referring to the manual for instructions on tuning up a lawn mower engine, he or she is probably in a garage or toolshed (with not enough light) or outdoors in bright sunlight (with too much light). Rarely are manuals read by someone sitting at a desk with a good reading light. In addition, the manual may see service in dusty or greasy environments (farm machinery manuals), be used by people with overworked eyesight (software manuals), or be used in a noisy environment that makes concentration difficult (industrial machinery manuals).

A good manual design makes it possible for the user to get the needed information despite these obstacles. Let's look now at how design elements can work together to make a user-friendly manual.

DESIGN CHOICES: WHY COOKBOOKS WORK BETTER THAN NOVELS

At the beginning of this section, we said that hardly anyone sits down and reads a whole manual through from beginning to end, as one would a novel. Instead, people use manuals more like cookbooks, opening them primarily to look up particular information. If I want to know how to make a chocolate cake, I don't want to have to read through an explanation of basic cooking techniques, how to equip a kitchen, guidelines for good nutrition, and the history of desserts in the Western world before I find the cake recipe. I want to be able to find the instructions for making a chocolate cake quickly and use them without having to wade through a sea of unneeded information. Once I find that recipe, I am likely to refer to it as I am actually making the cake, so that instead of reading it with rapt absorption as I might a novel, I will probably be looking back and forth between book and mixing bowl.

Manuals must be designed to make access to information easy for the user. An accessible manual has three key attributes:

- Information is *easy to find.*
- The manual *looks easy to read.*
- The manual is physically *convenient to use.*

In addition, the manual must be easy for the company to revise and adapt as needed.

Several elements must work together to make a manual in which information is easy to find. Clearly, the most crucial of these is the organization of the manual. Information must be placed in a logical order, so that the reader can navigate easily from topic to topic. Organization is covered in Chapter 3, but page design also plays a role. A good format can help reveal the organization to the reader, making it much easier and quicker to find what is needed.

The manual must look easy to read because it is in competition with all the other things we *should* read, but do not have time for. If the manual looks too hard, people simply won't read it; they'll spend their scarce time on something that looks more inviting. Whether a manual looks easy or difficult is largely a matter of whether the design is simple and straightforward or complex and confusing.

Finally, the manual must be convenient to use because readers will be using it in conjunction with the product, whether performing service or learning to operate it. If the manual is inconvenient, they will simply try to get along without it.

Making Information Easy to Find

Most people will leaf through a manual section before actually studying it in detail. What they are doing is trying to get a sense of how the section is put together — what the major divisions are, how the smaller pieces relate to the whole. The primary means at the reader's disposal to do this is the headings in the manual. Headings are like signposts that direct the reader to the appropriate part of the chapter. For the heading structure to work the way it is supposed to, it must accurately reflect the organization of the book and it must be detailed enough to be useful.

Heading Structure

The best way to generate a heading structure that is an accurate reflection of a text is to use the outline from which the text was written. An outline, after all, is simply a "sketch" of a piece of writing: it identifies the sequence and hierarchy of information. In other words, it tells what comes first, what second, and so on, and it tells how those sections are arranged into major and minor divisions. If the outline itself is logical, then a heading structure based on it will be logical as well.

Be sure to use enough headings. Most writers use too few. Instead of just labeling the major sections of a chapter, consider adding subheadings to point out the smaller divisions. Remember that, as the writer, you are familiar with what is contained in the manual you are writing — you know where the parts are and what is covered in each. Your readers are looking at the manual for the first time; without sufficient headings to help them find their way, the manual will look like a bewildering sea of prose.

On the other hand, be sure that your heading structure is not overly complex. Limit yourself to no more than three levels of headings — more than three becomes too confusing for the reader. Make sure that the type size and style that you have selected for the different levels of headings signal the shifts in an intuitive way. For example, we automatically assume that larger type identifies a major section, smaller type a subsection. Similarly, bold type of a certain size signals a more major division than normal type of the same size. A straightforward, intuitive heading design will help guide your reader through your manual without the reader even being aware of it. To illustrate how this works, look at Figure 4.1. It shows a page of Japanese text. Without being able to read Japanese, or even knowing the topic, we can still identify something about the structure — solely by looking at the headings.

A page designer then has at least three elements to work with in designing headings: type size, type style, and position. In a good design, these elements work in concert to guide the reader. In a poor design, they may be at odds and leave the reader frustrated and confused. The best way to ensure that your heading structure is consistent and rational is to make a simple table. You can then use the table to make sure that you have been consistent throughout the manual.

How do you decide what your headings should be? The simplest procedure is to use the outline for the text to guide you. Outlines show both the sequence of information and the hierarchy. Using the outline as a guide for the heading structure will ensure that your headings consistently and accurately reflect the organization of the material. The publication software your company uses may allow you to generate appropriately formatted headings automatically from an outline. The low-tech method of going through your outline with different colors of highlighter pens for the different levels of headings works just as well.

警告

1)指や手を入れないでください。けがをする恐れがあります。
2)お子様やペットを本機に近付けないでください。けがをする恐れがあります。
3)身につけている装飾品(ネックレスなど)やネクタイ、髪の毛等、本製品ご使用時に巻き込まれないよう、十分ご注意ください。
4)エアゾール式のクリーナーや潤滑剤など可燃性の製品をマシン内部や、本機の周辺に置かないでください。発火する恐れがあり、大変に危険です。
5)通常の操作以外に本機を移動させたり、掃除をしたりする場合は必ずメインスイッチを OFF にし、電源プラグをコンセントから抜いてください。
6)長時間の連続使用はさけ、適正裁断枚数を超える紙を入れないようにしてください。連続用紙、連続記帳をシュレッドしないでください。
7)このパーソナルシュレッダーは紙専用のシュレッダーです。カッター部分については5年間の保証をしており、またホッチキスの芯やペーパークリップをシュレッドしても、カッティングシステムを傷めることはありませんが、詳しいシュレッダーの操作につきましては、本取扱説明書の指示に従って行なってください。

操作方法 シュレッドの仕方

パーソナルシュレッダーPS 60はA4 サイズコピー用紙 8 枚、PS60CC は A4 サイズコピー用紙 7 枚をシュレッドすることができます。

1)メインスイッチをオンにします。
2)シュレッドする紙を用意します。ホッチキスの芯やペーパークリップなどはカッティング システムを傷めることはありませんが、シュレッドする紙の厚みが増しますので、その分シュレッドする枚数が減ります。(適正裁断枚数は製品仕様の項目をご覧ください。)
3)紙を垂直にペーパー投入口に差し込みますと、シュレッドシステムが自動的に作動し、シュレッドが済むと自動的に停止します。

注意:紙の幅が投入口よりも広い紙を折りたたんで投入した場合は、紙の厚み増しますので、その分シュレッドする枚数が減ります。

Figure 4.1 Selection of Japanese text. This selection of Japanese text from a multilingual manual for a paper shredder illustrates that even if we cannot read the text, we can still get clues about how the manual is organized by looking at the page design. (From *Operating Instructions, Fellowes Powershred Personal Series,* Fellowes Manufacturing Company, Itasca, IL, 1998. With permission.)

Headers and Footers

Another sort of signpost that readers can use to find their way to the information needed is the headers and footers at the top and bottom of the pages. Headers (also known as "running heads") are headings at the top of a page that tell readers what part of the book they are in. For example, it is common to show the chapter title at the top of the left-hand page and a subhead from within the text on the right-hand page. Less commonly, this information is given at the bottom of the page instead of the top, in which case the labels are called footers (not, to our knowledge, "running feet"). More commonly, footers, if used at all, are used for such information as revision number, series number, and so on. Running heads allow the user to flip through the pages quickly to find the relevant section, and then shift to using headings within the text to locate the precise information needed.

Table of Contents

Besides the signposts internal to the text, most manuals more than a page or two long have a table of contents. The table of contents is placed at the front of the manual and essentially provides a map of the book. It outlines what the manual covers, giving the starting page number for each section listed. To be useful, the table of contents must be arranged so that it clearly shows the organization of the manual and is easy to use. The names used to designate sections in the table of contents should match *exactly* the headings used in the text. You may not want to include all levels of headings in the table of contents, but those you do include should have the same wording in both places. You should also show organizational levels in the table of contents, perhaps by type size or indention, to reflect the arrangement of information in the text. Be careful to make it readable. If the page numbers on the right are too far away from the words on the left, the reader may find it difficult to know what goes with what. You may wish to run dot leaders across the page or leave spaces between small groups of headings and page numbers.

You must strike a balance between making the table of contents too skimpy and making it so complete that it nearly reproduces all the information in the manual itself. A good guideline is that if you find yourself repeating the same page number two or three times (indicating that the table contains multiple headings from the same page), your table of contents is probably too detailed. On the other hand, if the page numbers are 10 or 20 pages apart, the table of contents will not be much help in locating information. Of course, the level of detail in the table of contents is somewhat dependent on the overall length of the manual. One would not expect to see an entry for every three or four pages of a 500-page manual. Here again, using the outline will be helpful. At the same time that you decide on your heading structure, you can also decide what levels of headings will be included in the table of contents. Normally at least the first two levels would be included.

Some very long manuals may have a rather general table of contents at the beginning of the book and then a separate table of contents at the beginning of each chapter. These individual tables of contents provide a greater level of detail than the initial one. As an added help to the user, these manuals may set off individual chapters by color coding or reference tabs.

Index

The other comprehensive resource for a user trying to locate specific information in a manual is the index. The index is an alphabetical listing of subjects with the numbers of the pages on which the subjects appear. It is usually placed at the back of the manual and is much more detailed than the table of contents. Because it is alphabetical, rather than sequential, it does not reflect the organization of the manual.

Preparation of an index used to be a tedious process of putting each entry on a separate index card and going through the text writing down page numbers as each entry appeared. The use of computers has made it somewhat easier, since the computer can readily search for and locate words. However, developing a good index, one that will actually help users find what they need, is never a mechanical process. A good index is not just a list of *words* that appear in the text, but rather a list of *concepts*. The computer can help make sure that you do not miss any occurrences of a particular word, but a human still needs to make the decisions about what words to include. Keep in mind that the purpose of an index is to help readers find information: include words and concepts that you think someone might want to look up. And do not limit the reader to one approach: index the same concept with different wording. Think about looking in the Yellow Pages to find someone to service an outboard motor. Do you look under "Motor"? "Outboard"? "Boats"? "Engine Repair"? A good index will get you to the right destination from more than one starting place.

Cross-References

Equally important is the cross-referencing contained within the text itself. Often a manual user needs to be aware of information in a section other than the one presently in use. For example, in a lawn mower manual, the section on winter storage of the mower might say to drain the oil and replace it before the next use. The location of the drain plug and the proper weight of oil to use are listed in the section on maintenance. Rather than let the reader page through the manual at random looking for this information, point it out: "See 'Maintenance,' page 4." Put yourself in the reader's shoes and remember that he or she does not know the contents of the manual backward and forward as you do. Whenever you think it would be helpful for the reader to be referred to another section of the manual, do so. Of course, the most obvious place for cross-referencing is the troubleshooting section. This is usually set up as a table with headings like "Problem," "Probable Cause," and "Remedy." Too often the remedies suggest something like "adjust spark plug gap" without telling the user where in the manual the proper gap is given. To save your readers a lot of frustration, always include the full information needed — title of section *and* page number. If they are using the troubleshooting section, they are frustrated enough already.

Of course, providing specific cross-references means that those references have to be changed whenever the manual is revised and pagination changes. For this reason, some companies direct their writers to write "generically," omitting specific references — sometimes even to particular models of the product. As with many aspects of manual design, what makes the manual easier for the reader to use makes it harder or more expensive for the company. Each company must decide where the appropriate trade-off lies. Computer technology can help. Publishing software can keep track of page changes on flagged items, so that cross-references can be automatically updated.

One exception to the general rule that you should refer the user elsewhere in the manual for needed information is safety warnings. Do not refer users to safety warnings on pages other than the one being used. Users will simply not make the effort to turn the page to look at a warning. Instead, repeat the warning as many times as needed to make sure it is present when the reader needs to know about it. Remember, cookbooks, not novels. See Chapter 6, "Safety Warnings," for more information.

Numbering Systems

Some manuals number each paragraph. Various systems are used; perhaps the most familiar is that used in the military. In this system, hierarchy is indicated by numbers divided by periods. Thus, paragraph 1.2 is of a higher level than paragraph 1.2.1. In outline form, it might look like this:

1.0 Chapter One
    1.1 First Subtopic
    1.2 Second Subtopic
        1.2.1 First Sub-Subtopic
        1.2.2 Second Sub-Subtopic
2.0 Chapter Two ... and so on.

The advantages of such a system are that it is easy for the reader to see the organizational structure of a manual and that it is possible to direct a reader to a particular paragraph (rather than just a page) for specific information. Further, if a company has a number of similar products, common, "boilerplate" information may be easily catalogued on the computer by giving each paragraph a unique address, making it possible to assemble relevant information quickly and easily.

The system has some disadvantages as well, however. It is a bit of a distraction to the reader, adding a level of clutter to page layout that may or may not outweigh the usefulness. Whenever the manual is revised, even in a minor way, the paragraphs have to be renumbered, particularly if information is inserted rather than deleted. It may get quite cumbersome if there are many levels to the outline. Finally, it may be confusing to the reader when combined with writing strategies such as using numbered lists for explaining procedures.

Whatever system is used, it must show hierarchy as well as sequence. Some years ago, certain United Nations documents used a purely sequential numbering system: paragraphs would be numbered in order from the beginning to the end of the document. Although such a system allows references to particular paragraphs, it does not help the reader see the structure at all, and *any* revision requires renumbering the entire document.

When a paragraph numbering system is used, the index (and sometimes the table of contents) generally refers the reader to the paragraph number rather than the page number associated with the listed topic. As long as this is clearly indicated, it should pose no problem, although it is a little more difficult to flip through a manual to find a specific paragraph number than to find a specific page number — simply because the page numbers are always in the same location on the page.

Making the Manual Look Easy to Read

Have you ever picked up a manual and almost immediately put it back down, thinking, "That looks too hard to read"? Often, first impressions will determine whether all your hard work in writing a manual will be time well spent or an effort all for naught. It is difficult enough to get people to read product manuals — we do not want to put them off further because of poor page design. We have covered some of the aspects of page design that make a manual useful to the reader: such things as a functional heading structure, sensible reference information, and so on, but what makes the page visually appealing? If users never get past a negative first impression, they will never be able to appreciate the well-thought-out heading structure. A page design that looks inviting and readable is one that uses a type size and style that are easy to read, visually chunk related information, and include adequate white space. Let's look at those elements more closely.

Type Size and Style

Typefaces come in a bewildering variety of styles and sizes, but all English typefaces fall into two broad groups: serif and sans serif. Serifs are the little lines at the end of each stroke in a letter. Serif type has these lines, and sans serif type does not (*sans* is French for "without"). Figure 4.2 shows samples of serif and sans serif type.

Sans serif type has a "modern" look to it, but most people perceive it as being harder to read. No one is quite sure why this is — it may be that the serifs tend to "seat" the letters on the line and pull the eye along to the next word. Although sans serif type *seems* harder to read, studies show both styles perform well in actual tests of reader comprehension. Nevertheless, because of the eye appeal of serif type, we recommend using it for the main text and reserving sans serif styles for headings.

Similarly, a mixture of uppercase and lowercase letters (as in ordinary text) is easier to read than text written in all uppercase. This difference is not mere perception, however. Lowercase letters show much more variation in their shapes than do uppercase letters — all of which are variations on circles, rectangles, and triangles. Thus, it is easier for the eye to distinguish one letter from another in lowercase. A word or two (such as in a heading) in all uppercase is fine, but never set a whole paragraph in uppercase type. Sometimes well-meaning page designers decide that setting the message block of a safety warning in all uppercase letters is a good way to emphasize it. Unfortunately, it has the opposite of the desired result: people will be less likely to read the warning than if it were in ordinary mixed case.

More important than any other aspect, however, in whether a typeface looks inviting and readable is whether it is big enough. Type is measured in *points*, with

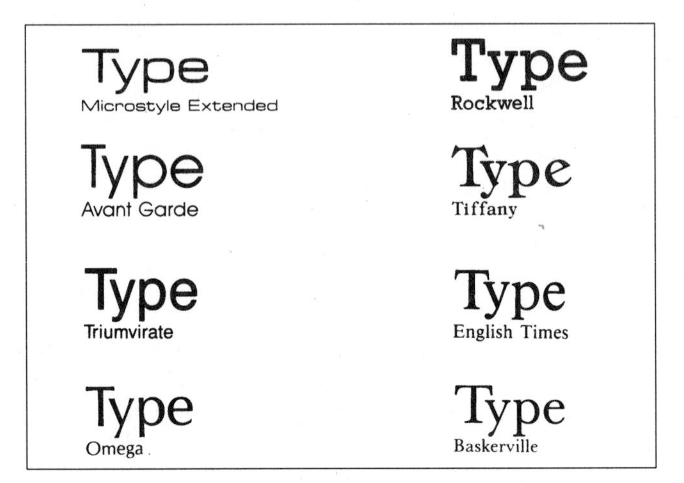

Figure 4.2 Examples of serif (right) and sans serif (left) type. Sans serif typefaces normally use strokes of single weight, or width. The Omega typeface shown here is unusual, in that the lines widen out toward the ends — almost suggesting the beginnings of serifs.

72 points to the inch. Before computers, which now do it all electronically, type was printed mechanically. Each line of type was manually set by a typesetter, who aligned individual metal blocks with raised letters on them to spell out a line of type. Point size refers not to the actual letter size, but rather to the height of the metal block on which the raised metal letter sat. Thus, in two different typefaces, the same 12-point letter might be slightly different heights, but the blocks would be the same. Even though hand-typesetting has been made obsolete by the computer, the old terms are still used.

With 72 points to an inch, the tallest 12-point letter would be be a little less than 1/6 of an inch high. The smallest size that can be read without a magnifying glass is 6-point type. Most text is set at least in 8-point type — which still requires reading glasses for those of us over 40. Because manuals may well be used in less than ideal conditions, we recommend going no smaller than 8-point type for anything in the manual and would urge using 10-point or 12-point type for ease of reading. Figure 4.3 shows how point size can affect readability.

Visual Chunking

From looking at Figure 4.3, it is obvious that type size also affects line length. The same sentence set in 14-point type is significantly longer than one set in 8-point type. The best line length is around 40 to 50 characters. With a line that is very much shorter than that, the eye must keep jumping to the next line, and words and phrases are frequently broken, causing fatigue and decreased comprehension. If the line is much longer, the eye has too far to travel back to begin the next line and is likely to settle on the wrong line of type. (A corollary is that the longer the line, the more space is needed between lines.) While a longer line length may be acceptable

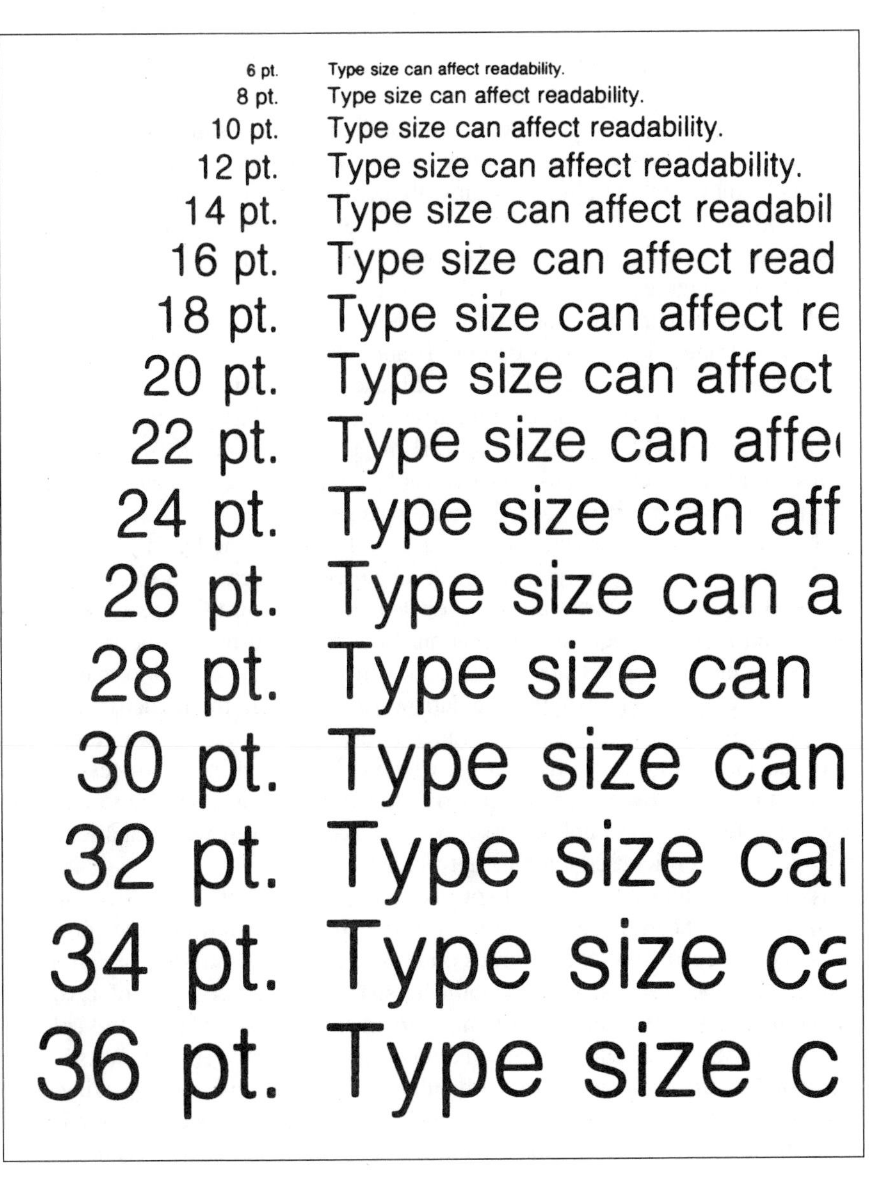

Figure 4.3 Samples of different type sizes, measured in points. For ease of reading, the text in manuals should be at least 10-point type. Poor working environments indicate the need for even larger type sizes.

in books and reports, where the reader is expected to read through from beginning to end, it is not a good idea in manuals, where the reader is likely to be looking back and forth between manual and product. A line length closer to the optimum works best.

Many manuals use an 8½ × 11-inch page size (more or less). Text set in 10-point type running the full width of the page would be very difficult to read, so page designers have come up with various ways to shorten the line length, primarily by arranging the text in columns. The next set of figures shows some sample pages, all roughly 8½ × 11 inches in the original, with the text laid out in different ways. Figure 4.4 shows a two-column layout with the right margin of the text *ragged*, that is, not forming a straight line like the left side. Figure 4.5 shows another two-column layout, this time with the right margin *justified* (the edge of the text forms a straight vertical line). Figure 4.6 shows a three-column layout.

A variation of the two-column layout is the "2/5" format. In this format, the columns are uneven: the larger column is about 5 inches wide and is used for the text, while the narrower column is about 2 inches wide, and is used for illustrations or notes. On facing pages the column arrangement may be the same, or it may be reversed so that facing pages are mirror images of each other. Additionally, the narrow column, which may also be used for headings or other explanatory material, may fall at the inside or the outside of the page.

No one layout is perfect; all have trade-offs. For example, in a two-column format, if the columns are right-justified, the space between them can be narrower without the columns appearing too crowded. But right-justified text looks more formal and requires frequent hyphenation. In addition, justifying the right margin means that the spacing between letters and words must be adjusted to make all the lines come out the same length. The narrower the column, the harder this is to do well. We have all seen dreadful examples in newspaper columns where one long word has been s t r e t c h e d across a whole column, and in the process become nearly unreadable. Before you opt for right-justified margins, be sure to check out how sophisticated your publishing software is. Badly spaced words and letters can make text very difficult and unpleasant to read.

Whatever column arrangement you decide on, stick with it for the whole manual. It is quite disconcerting to the reader to turn a page and find that the format has suddenly switched from two columns on a page to three. Similarly, be careful that your illustrations do not protrude into the space between columns of text. They should be sized to fit within the column or to extend over two or more full columns. If the illustrations are allowed to "leak" into the gutter between columns, the page will look visually disorganized and difficult to read. Remember, the object in designing a format is to make the pages look inviting. A clean, consistent, ordered layout gives the impression that the material is well organized and easy to read.

Adequate White Space

An inviting page layout includes adequate white space to help the reader group related information and to provide physical rest to the reader's eyes. Sometimes writers are tempted (or pressured) to cram as much as possible onto each page in order to cut production costs — after all, less "empty" space means fewer pages. But eliminating needed white space to cut costs is a false economy: the resulting overcrowded layout will discourage even the most dedicated manual readers. Anyone else will not even try to wade through it.

CARBURETOR ADJUSTMENT

This unit is equipped with a diaphragm-type carburetor that has been carefully calibrated at the factory. In most cases, no further adjustment will be required.

The condition of the air filter is important to the operation of the trimmer. A dirty air filter will restrict the air flow, which upsets the fuel-air mixture in the carburetor. The resulting symptoms are often mistaken for an out-of-adjustment carburetor. Therefore, **check the condition of the air filter before adjusting the carburetor**. Refer to **Air Filter Maintenance**.

If the following conditions are experienced, it may be necessary to adjust the carburetor:

- The engine will not idle
- The engine hesitates or stalls on acceleration
- The loss of engine power that is not corrected by cleaning the air filter and muffler
- The engine operates in an erratic or fuel-rich condition (indicated by excessive exhaust smoke from the muffler).

☐ NOTE: Careless adjustments can seriously damage your unit.

Adjusting the Carburetor

1. Clean the air filter if it is dirty. Refer to Air Filter Maintenance.
2. **Initial Idle Speed Setting:** Turn the idle speed screw **counterclockwise** (Fig. 26) until it *does not contact* the carburetor throttle lever. Now turn the screw **clockwise** until it *begins to move* the throttle lever; then continue turning **2 full turns**.
3. **Initial High Speed and Idle Mixture Setting:** Turn both the high and idle mixture screws **clockwise** (Fig. 26) until they are *lightly* seated. Then turn the screws **counterclockwise 1-1/4 turns**.
4. Start the engine and let it run for a minute.
5. Release the throttle trigger and let the engine idle. If the engine stops, turn the idle speed screw (Fig. 26) **clockwise 1/8 turn** at a time (as required) until the engine idles.

☐ NOTE: Turn the high speed and idle mixture screws finger-tight. Forcing the mixture screws with a screwdriver will damage the screw tip and the seat in the carburetor body.

6. **Final Idle Speed and Mixture Settings:** Adjust the idle speed and mixture for smoothest engine idle.
   a. Turn the idle mixture screw (Fig. 26) clockwise until you hear the fastest idle; then turn the screw **counterclockwise 1/8 turn**.
   b. Squeeze the throttle trigger. If the engine falters or hesitates as it accelerates, turn the idle mixture screw (Fig. 26) **counter-clockwise 1/16 turn** at a time until the engine accelerates rapidly.
   c. If the idle speed changes significantly because of steps a and b, readjust the idle speed screw (refer to Step 2).

☐ NOTE: The Bump Head line spool should not rotate when the engine idles.

7. **High Speed Screw Mixture Adjustment:**
   a. High speed mixture screw adjustment is not recommended without a precision high speed tachometer.
   b. The factory presets the high speed mixture screw at 1-1/4 turns out from the closed position. Your unit should perform well at this setting. If additional adjustment of the high speed mixture is required, contact your local authorized service dealer.

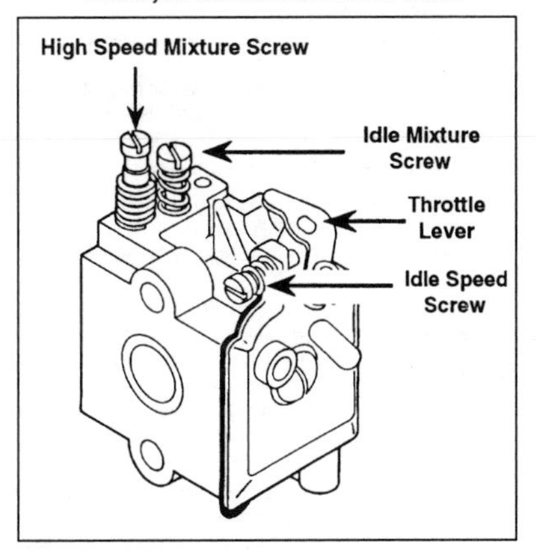

Fig. 26

☐ NOTE: If the carburetor adjustments do not help the unit to run properly, contact your authorized service dealer.

Figure 4.4 Two-column layout, set ragged right. In this example, from a manual for a lawn trimmer, the irregular, or "ragged" right margin on each column of text gives the manual an informal, open look. (From *Operator's Manual, IDC 500*, Ryobi Outdoor Products, Inc., Chandler, AZ, 1998, p. 11. With permission.)

White space can help the reader to understand the structure of a piece of writing. In the same way that headings can show how the manual is organized, white space, if used carefully and consistently, can show how a section of text is put together. Thus, a blank line or two (such as between paragraphs) lets your reader know that

Introduction

100 square inches), if each opening communicates with other unconfined areas inside the building. Buildings of unusually tight construction shall have the combustion
and ventilation air supplied from outdoors or a freely ventilated attic or crawl space.

If air is supplied from outdoors, directly or through vertical ducts, there must be two openings located as specified above and each must have a minimum net free area of not less than one square inch per 4000 BTUH of the total input rating of all the appliances in the enclosure.

If horizontal ducts are used to communicate with the outdoors, however, each opening must have a minimum net free area of not less than one square inch per 2000 BTUH of the total input rating of all the appliances in the enclosure. If ducts are used, the minimum dimension of rectangular air ducts shall be not less than 3 inches.

NOTE: If the openings are to be covered with a protective screen or grill, the net free area of the covering material must be used in determining the size of the openings, as stated above. Protective screening for the openings MUST NOT be smaller than 1/4 inch mesh to resist clogging by lint or other debris.

Provisions for combustion and ventilation air must comply with referenced codes and standards. See Local Installation Regulations Section on Page 4.

F. **CORROSIVE ATMOSPHERES**—The water heater should not be installed near an air supply containing halogenated hydrocarbons. For example, the air in beauty shops, drycleaning establishments, photo processing labs, and storage areas for liquid and powdered bleaches or swimpool chemicals often contain such hydrocarbons. The air there maybe safe to breathe, but when it passes through a gas flame, corrosive elements are released that will shorten the life of any gas burning appliance. Propellants from common spray cans or gas leaks from refrigeration equipment are highly corrosive after passing through a flame. The limited warranty is voided when failure of water heater is due to a corrosive atmosphere. (Reference is made to the limited warranty for complete terms and conditions.)

Installation

1. **INSPECT SHIPMENT**—Inspect water heater for possible shipping damage. Check the marking of the rating plate of the water heater to be certain the type of gas being furnished corresponds to that for which the water heater is equipped.
2. **WATER SUPPLY CONNECTIONS**— Refer to Fig. 2 for suggested typical installation. The installation of unions or flexible copper connectors is recommended on the HOT and COLD water lines, so that the water heater may be easily disconnected for servicing if necessary. The HOT and COLD water connections are clearly marked.

   Install a shut-off valve in the cold water line near water heater.

   Determine if there is a check valve in the cold water supply line. It may have been installed as a separate component or it may be part of a pressure reducing valve, water meter or water softener.

   A check valve located in the cold water inlet line can cause a "closed" water system. A closed system prevents the water, as it is being heated, from expanding back into the cold water supply line. Pressure can build up within the heater causing the relief valve to operate during a heating cycle. This excessive operation can cause premature failure of the relief valve and possibly the heater itself.

   Replacing the relief valve ***will not*** correct the problem. One method of preventing pressure build-up is to install an expansion tank in the cold water supply line between the heater and the check valve. Contact your installing contractor, water supplier or local plumbing inspector on how to control this situation.

   IMPORTANT!! This water heater is supplied with one (1) "Cold" (blue tip), and one (1) Hot (red tip) heat trap nipples. They MUST be installed directly into the water heater as shown in Fig. 2. Do not apply heat to the hot or cold heat rap nipple. If sweat connections are used, sweat tubing to

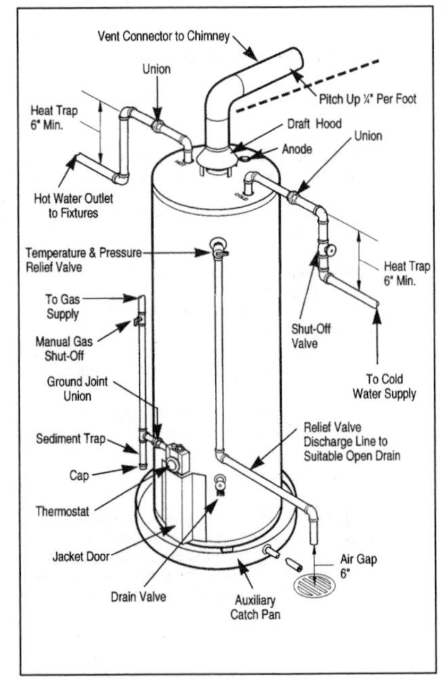

Figure 2 — Typical Installation

Figure 4.5 Two-column layout, set fully justified. In this example, from a manual for a water heater, the right margin is justified (straight), giving the page a neat, formal look. Notice, however, that even though the illustration on this page is larger than the one in Figure 4.4, this page appears to have more text on it. Using fully justified text requires taking care to ensure that each page has sufficient white space to provide rest for the eye and to aid in visual chunking of information. (From *Use & Care Manual, Tri-Power Residential Gas Water Heater*, Rheem Mfg. Co, Montgomery, AL, p. 2. With permission.)

USING THE BORDER EDGER ATTACHMENT

Your Mantis Tiller has been designed and built to accept a wide range of Mantis Tiller Attachments to increase its usefulness in your lawn and garden. And, all Mantis Tiller Attachments have been designed for quick and easy attachment to the Tiller or Engine.

The Border Edger

(Item #3222M)
The Most popular attachment, the Border Edger can be used to cut clean, neat edges along walkways, or around trees, shrubs, and garden beds.

The Border Edger has two parts: a wheel and a hardened steel blade, with pointed tines.

How to Install the Border Edger

The following instructions refer to "right" and "left" axles. Assume that you're standing behind your Tiller, as you would for tilling and cultivating.

Picture 1

Some areas of your yard may harbor roots and other underground obstructions. In places like this you'll want to edge your borders shallowly (1" to 2" deep). Here's how to install the Border Edger for shallow edging:

1. First remove your tilling/cultivating tines.

2. Then slide the Edger's wheel onto the right axle.

3. Now slide the Edger blade onto the left axle. The blade's angled face should hit the ground when you spin the blade forward.

4. Insert retaining pins on both left and right axles.

Around walkways and garden beds, you'll want to edge more deeply (3" to 4" deep). Here's how to install the Border Edger for that purpose:

1. Remove the tilling/-cultivating tines.

Picture 2

2. Slide the Edger's blade onto the right axle. The blade's pointed face should hit the ground when you spin the blade forward.

3. Slide the wheel onto the left axle.

4. Insert retaining pins on both sides.

How to Use the Border Edger

1. Position your Mantis Tiller so that the Edger blade is right along the garden edge and the wheel is outside (on the lawn, on the sidewalk, wherever). (Picture 1.)

2. Start your Tiller and pull your Mantis backward along the garden edge. (Picture 2.)

The Border Edger Can Handle Special Projects!

1. Install the Edger for deep edging, as directed above. Then use it to cut sod strips.

2. Edge and weed at the same time! Just attach the Edger blade on one axle and a Tiller tine on the other axle, "Mix and match" blades; don't be afraid to experiment.

Important Note: If you do a lot of edging, you'll appreciate the Mantis Wheel Set (Item #9222M.) It gives you added stability, for even easier handling.

To order the Wheel Set, or any Mantis Attachment, call 1-800-366-6268, Monday through Friday, 9 am to 9 pm, Eastern Time. Ask for the Sales Department.

Figure 4.6 Three-column layout. In this example, from a manual for a garden tiller, the text is set in three columns, providing an open look with a good deal of white space. With such a narrow column width, ragged right is a good choice — there's not much room to adjust word or letter spacing as is required for justified text. The column width is ample for the sorts of graphics used here, but would be too small for detailed assembly drawings or charts.

you are moving from one unit to another of equal importance. Similarly, indenting a section (widening the white space around it) lets your reader know that you are moving to a smaller organizational division within a single unit. Surrounding an item with white space will also call attention to it, such as setting off a warning from ordinary text for emphasis.

Properly used white space can also make it physically easier to read a page. Our eyes, like the rest of our bodies, need rest breaks. If these mini-rests are not provided to us in the form of white space, our eyes will take them anyway — and we will find ourselves having to reread a sentence because we missed a few words. Also, most people are able to read faster than their minds can follow. White space gives the mind time to assimilate a piece of information before going on to the next.

As with any other design element, white space must be used consistently to work properly. In other words, you must always use the same number of blank lines between major divisions in a manual, and the same (smaller) number between minor divisions. As with headings, you will find it easiest to keep track of this if you make yourself a table. Otherwise, you will be having to page or scroll back through completed material to find whether to leave two lines or three between sections.

Careful use of white space may add a tiny bit to the cost of a manual because not every page is crammed corner to corner, but it will help ensure that the manual is used. If the manual sits untouched on the shelf because it looks too hard to read, the entire cost of producing it is wasted.

Making the Manual Physically Easy to Use

Design choices that need to be made include more than just page design — and some of those have implications for page design. For example, what is the appropriate page size? Is the manual going to include different languages? Will it need to be updated frequently? What kind of paper should it be printed on? All these decisions will affect how easy it is for the customer to use the manual and how easy it is for the company to revise it.

Size

The appropriate size for a manual depends on how and where it will be used. An 8½ × 11-inch three-ring binder would be an awkward size for a manual for a 35-mm camera or a handheld GPS unit — you want something small to fit in the case. On the other hand, it would be a fine size to use on a workbench in a garage. Think about how your customer will want to use the manual. Also consider how the company would like to have the manual used. It may be important in reducing liability exposure to have it convenient to keep the manual with the product. Sizing an operator manual to fit in a compartment built into an industrial machine might help ensure that the machine operator has access to the manual. On the other hand, the three-ring binder with letter-size pages is the standard for industrial reference manuals. In general, it is best to use common sizes: oversized manuals won't fit in a bookcase and small, odd-sized manuals are easy to lose.

Paper

The paper used in a manual must be durable enough to last the life of the book. If a manual is likely to be read once and discarded, the paper need not be of the best quality. On the other hand, if the manual will be referred to again and again, the paper needs to be durable. Another consideration is that the flimsier the paper, the more likely it is that the printing on one side of the page will *bleed through*, or be visible from the other side. Bleed-through makes reading more difficult.

The paper used for a manual must also not be too porous. Porosity affects how the paper accepts ink and thus how sharply photographs and illustrations will appear. If you have ever tried to write on a paper napkin with a fountain pen, you have experienced the tendency of porous paper to allow ink to bleed. If the ink can bleed, lines become fuzzy, and fine detail is lost. Porous paper will also accept other substances, like oil and dirt. If your manual is likely to be used under dirty conditions, you should choose a harder-surfaced paper, possibly even a coated stock (although this is quite expensive).

Binding

A good binding holds the pages of a manual together so that they can be easily read. Here are some factors to consider:

1. *Will the pages lie flat?* Nothing is more irritating than trying to do a procedure requiring both hands and frequent glances at the instructions, only to find that the manual flops shut each time you let go.
2. *Will the pages begin to fall out after hard use?* This is a common problem with "perfect" (glued) bindings, although rarely with stapled, stitched, or spiral bindings.
3. *Will frequent additions or corrections be sent to owners?* If the manual is likely to be updated often, it might be a good idea to put it in a ring binder, so that outdated pages can easily be replaced with new ones. If the binding requires that holes be punched in pages, make sure that margins are wide enough to allow this to happen without losing part of the text. An alternative for occasional changes is to send adhesive-backed pages to owners. These pages can be stuck onto the existing page in a bound manual. It is certainly a more expensive option, but for some vital information it might be worth the cost.

Cover

The cover of a manual influences both how likely it is that the manual will be used and how easy it is to use. An attractive cover invites readers and presents a good image of the company. Remember, the manual offers an avenue for the company to show its customers that it cares about them. An attractive manual cover is the first step. Most companies have a standard cover format for their manuals, including information about the model and often a picture of the product. A good cover should also include the company's address and telephone or fax number (the back cover is fine for this).

The cover of a manual must also be stout enough to protect the inside pages under the expected conditions of use. Clearly, if the manual will be used in a harsh environment (rain, dirt, oil, etc.), it should be made of coated stock, and possibly even of vinyl. On the other hand, a manual that will receive only occasional use in a clean office may have a cover of the same paper as the inside pages.

Making the Manual Easy for the Company

While it is crucial to design a manual that users find appealing, it is also important to design a manual that works well for the company. In today's business climate of fierce competition and a global marketplace, manuals must be flexible, adaptable documents. They must be easy to update when new information becomes available or products are changed; they must be easy to translate for overseas markets; and they must be easy to combine into new manuals for new products. In some cases, they may need to appear as online documents linked to a company's Web page. Some of the characteristics of manuals that meet these varied requirements are as follows.

- *They follow a consistent, company-wide format.*
- *They are organized in stand-alone modules.*
- *They depend more on visuals and less on text.*

As we shall see, these characteristics may require that you rethink how a manual is developed, but they will in the end make it more useful. Let's look at them more closely.

Consistent Format

You should develop a good format and use it for all the manuals your company produces (at least within a given product line). A standard, consistent look will help with readability, especially for repeat customers, but it will also facilitate reusing parts of a manual in a revised version or in a manual for a related product. Just because your company has a company standard in place, do not assume that it is ideal. Take a fresh look at the standard format in light of the considerations addressed in this book: Is it time for a new look? If you do decide to change the format, make your design choices carefully, so that the new standard will serve effectively for a long time.

Modular Organization

Whenever possible, organize information within manuals as modules, perhaps with associated illustrations or photographs (see also Chapter 3). These modules should be as close as possible to stand-alone pieces of information. The easiest way to think of this is to design in terms of a two-page spread — which is, after all, what is visible at one time to a user of a traditional manual. (Of course, for an online manual, the process is a bit different: there the "page" is a screen, but the principle remains the same.)

If you can design a stand-alone module for a particular procedure, such as adjusting the choke on a two-cycle engine, you can use that same module for several manuals dealing with different models of engines. You don't need to start over each time. These modules can be stored as computer files, allowing a new manual to be "built" from components of previous ones. Only information specific to the new product would have to be developed.

Visual Basis

The emergence in the second half of the 20th century of the global marketplace has created both opportunities and problems. One of the problems is finding a way to convey needed instructions and warnings in a variety of languages. Lots of approaches have been tried, from using a controlled English vocabulary for manuals to having them translated. Chapter 8 discusses these in more detail. In either case, generally it is the text that needs to be translated, not the visuals. Photographs and illustrations are meaningful whether the user speaks French, Japanese, Swahili, or English. The captions and callouts may need to be translated, but the image itself does not. Therefore, manuals that are built around visuals will be easier and cheaper to make usable for international audiences, solely because there is less text to translate. Naturally, some products lend themselves more readily to this approach than do others. In any case, developing visual-based manuals requires careful planning to make sure that the visuals really do carry the meaning intended.

SUMMARY

We began this chapter with a discussion of how technology has changed the art of manual production and made the writer responsible for many more aspects of a manual than just the writing. This challenge brings with it the opportunity to design the manual as an integrated unit that makes it easy for a user to locate needed information. More attention to the design aspects of manuals comes as a natural outgrowth of the way manuals are used in today's world: by people with too little time and too many competing demands. Careful choices in designing layouts and choosing "packaging" for a manual will pay dividends in attracting users to make use of the manual and benefit from its instructions and warnings.

At the same time, good design choices will make it easier for the company to adapt manuals to new products and new markets. Today's marketplace is the whole world, and manufacturers must compete not only on the basis of their products, but also the documentation that goes with them. A user-friendly, flexible, modular manual will work as well in overseas markets as at home. Good documentation encourages repeat business as customers respond to a company that clearly has their interests in mind.

The issues this chapter addresses, such as heading structure, cross-references, type size, column width, and paper, are unnoticed by most readers — unless they are not done well. This is as it should be: good technical writing is transparent; the reader sees right through the medium to the message. But to achieve that transparency requires careful attention to a host of design details — details that often make the difference between a manual that sees repeated use and one that is thrown down in frustration.

CHECKLIST

- ☐ Is the cover of the manual attractive and informative?
- ☐ Is the manual a convenient size?
- ☐ Will the binding stand up to the conditions of use?
- ☐ Is the page layout attractive — not too crowded or jumbled? Does it look organized and purposeful, rather than "thrown together"? Is there enough white space?
- ☐ Have I created a useful and consistent heading structure that reveals the manual's organization?
- ☐ Does the manual have headers and footers that help locate information?
- ☐ Is the print big enough to read easily?
- ☐ Are there cross-references to other information in the manual?
- ☐ Have I included a table of contents and index?
- ☐ Will the manual be easy to revise and/or translate?
- ☐ Does the design of the manual invite the reader in? Does it give the impression that the manual is clearly written and that information will be easy to locate?

FURTHER READING

Here are a few books that can help you with page design and layout. They are not specifically directed toward the design of manuals, but rather address the broader principles of good design. For further help, search the trade journals, such as *Technical Communication.*

Baird, R. N., with McDonald, D. *The Graphics of Communication: Methods, Media, and Technology*, 6th edition, Harcourt Brace College Publishers, New York, 1993.

Craig, J. and Meyer, S. E. (Eds.). *Designing with Type: A Basic Course in Typography*, Watson-Guptill, New York, 1999.

Schriver, Karen A. *Dynamics in Document Design*, John Wiley & Sons, New York, 1997.

Siebert, L. and Ballard, L. *Making a Good Layout*, North, 1992.

CHAPTER 5

Choosing and Designing Graphics

OVERVIEW

For most users, the graphics in a manual play the biggest part in creating a first impression. As one leafs through the pages, it is the graphics — photographs, drawings, charts, and tables — that make the manual look inviting ... or forbidding. Sharp, clear photographs and illustrations will tell the user that the company cares enough about the customer to produce a top-quality manual. On the other hand, fuzzy pictures and cluttered drawings leave the reader with the feeling that the manual was an afterthought.

Graphics do more than make an impression. They are a vital part of the communication package as well. In our increasingly visually oriented society, they may become the primary communication channel for some purposes and some audiences. Graphics work better than words to help a user identify parts on a product or learn to do a procedure. For readers who do not read well or who do not know English, graphics may be the the only source of information about the product. Even in a translated manual, the graphics remain critical — they are not subject to the vagaries of poor translation. Indeed, because they don't *need* to be translated, including graphics will reduce the cost of producing manuals for foreign markets.

To provide all these benefits, graphics need to be designed and executed well. This chapter provides an introduction to effective use and design of graphics in manuals. The best graphics result when manual designers plan them from the very beginning, right along with the text. When graphics are a part of the plan from the start, they can be integrated seamlessly with the written portions so that text and art work together to convey the needed information to the user. Good design of graphics follows the same basic tenets of good textual design: they must be easy to use and must make important information stand out.

WHY USE GRAPHICS?

Elsewhere in this book, we have discussed how to plan and write effective instructions. Well-written descriptions of products and explanations of procedures are important ways to convey needed information to users. But they are not the only way to do so — and sometimes not even the best way. In manuals, the graphical elements, such as photographs, drawings, charts, and tables, may be more important than the words. Clear, readable instructions and descriptions are necessary, but clear visuals are vital.

Why are graphics so important? Graphics have three key attributes that make them a critical part of manual design: in any manual, well-designed graphics

1. Attract the eye
2. Convey some kinds of information better than prose
3. Facilitate translation

Each of these makes it more likely that a manual will be used and understood — which in turn makes it more likely that the product will be used properly and safely.

Graphics Attract the Eye

As a society, we are less and less word oriented, and more and more picture oriented. The first generation to grow up watching TV is now middle-aged. The Internet, with its graphics-based Web pages, has gone from being used mostly for entertainment to being used for serious research and commerce. Every hit song has an accompanying video. No one in business would consider making a presentation without slides, transparencies, or a PowerPoint™* show. We have come to rely on graphical presentations of information nearly as much as on verbal.

Most users, confronted with a manual for the first time, will leaf through the "pictures" before they will read the words. Most people will try to figure out a procedure from the pictures first and will read the text only as a last resort. Readers will not only see the pictures before the text; they will usually expect them to be more accessible than the text. Graphics can break up large blocks of text and give the manual a more open, "friendly" feel. By contrast, a manual that is only text looks extremely forbidding to most readers, even if the font is large enough for easy reading and there is adequate white space.

Graphics Convey Information Better

Some things are simply easier to grasp when they are presented graphically rather than in words. For manuals, graphics are especially important for these functions:

- Identifying parts of a product
- Describing certain procedures
- Presenting quantitative information
- Explaining spatial relationships

Each of these functions lends itself much better to pictures than to words. Let's see why.

* PowerPoint is a trademark of Microsoft, Inc.

Identifying Parts

All manual users will rely on the pictures to help make unfamiliar part names clear: when told to "tap leg closures firmly until well seated," even the most word-oriented owner of a new barbecue grill will look at the drawing to find (with relief) that the leg closures are merely the little plastic caps that go over the ends of the grill's tubular metal legs. Graphics are used for parts identification throughout a manual, from the initial setup chapter to the parts catalog at the end. Words absolutely cannot substitute for good visuals in helping the reader to get to know the product. Figure 5.1 shows a typical use of a graphic for parts identification.

Describing Procedures

A standard classroom exercise to illustrate the importance of clear instructions is to have one student give verbal instructions to another on how to tie a necktie or make a peanut butter sandwich. The exercise is set up so that the student giving the instructions cannot see the student carrying them out. The rest of the class, however, can, and the result is always very entertaining. It points up the fact that sometimes it's far easier to *show* someone how to do something than it is to *tell* how to do it.

In manuals, the photographs and drawings do the "showing." With a graphic, a writer can show a user where to place a tool, how to test a belt for proper tension, or how to identify a hose that is ready for replacement. A series of pictures can lay out a sequence of actions, almost as if the user had an expert-in-residence to demonstrate the procedure in person.

Presenting Quantitative Information

Manuals typically contain a good deal of quantitative information. This information can range from gear ratios to baud rates, from load limits to dip switch settings. Generally, quantitative information is much more effectively conveyed in the form of a table, chart, or graph than in words. This is especially true if the reader needs to be able to compare the quantities or otherwise relate the information. Numbers buried in paragraphs of prose are easy to miss and hard to relate to one another. Just as putting instructions in a series of numbered steps makes it easy for the reader to follow, putting information that the reader must relate (such as operating conditions or frequency of lubrication) in a table or chart makes it easy to see the connections.

The connections become obvious because charts and tables, like drawings and photographs, can be seen as a whole. Closer inspection may be needed to see all the details, but the entire graphic can be seen all at once. Prose, on the other hand, is essentially linear: you can skim it, but cannot see it all at once. The information is presented sequentially. As a consequence, paragraphs full of numbers become very confusing, forcing the reader to try to remember all the numbers until the relationships have been explained. It is particularly difficult if different kinds of numerical information are presented — percentages and costs, for example — because typically like units are not next to each other for easy comparison, but rather in separate sentences. Such information is far easier to grasp when presented in a table or graph.

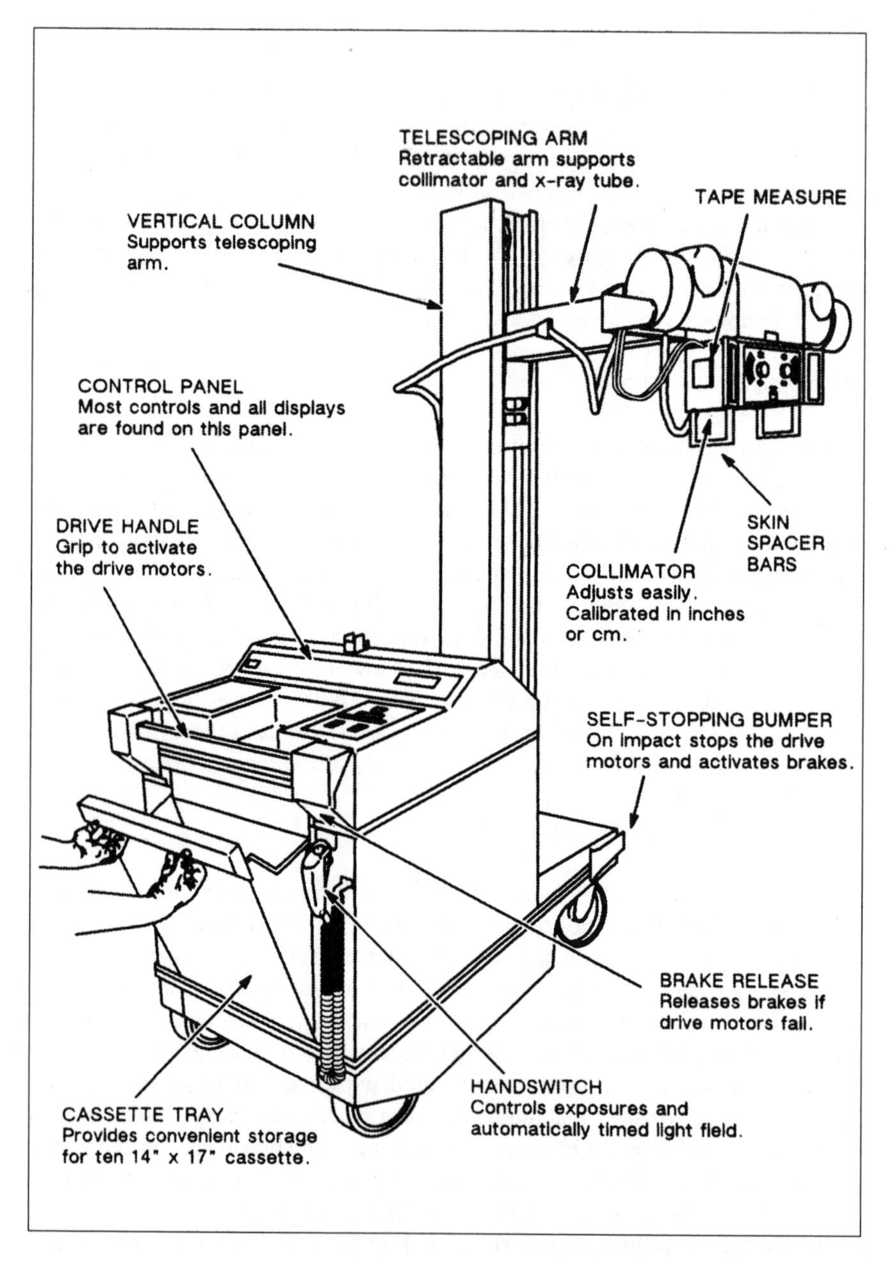

Figure 5.1 Typical use of a graphic to identify parts. (From *Direction 17291 Revision D, AMX 4 Operation*, General Electric Medical Systems, Milwaukee, WI, 1989, 4. With permission.)

Explaining Spatial Relationships

As with procedures, it is far easier to *show* where two parts of a product are in relation to each other than to describe the relationship in words. Graphics can become especially important when a description depends on the product being oriented in space in a particular way. What constitutes the "left" side, for example, changes with where one is standing in relation to the product. A good graphic can save many words of explanation.

Graphics Facilitate Translation

The pictures, or graphical elements, take on even more importance when the manual is written for an international market. Selling products overseas (or even in Canada) commonly means producing the manual in more than one language. Well-designed and well-reproduced graphics can make translation more cost-effective for international manuals. The reason is that English text tends to expand when translated into any other language. English has a very rich vocabulary, with many near-synonyms, differing only in nuance or connotation. Thus, one can be very precise (by choosing just the right word) and, at the same time, very concise (since "just the right word" exists). Other languages, with a smaller lexicon of words to choose from, require more words to attain the same precision. Thus, producing foreign-language text usually means higher printing costs on top of translation costs. "Pictures" do not need to be translated like text (although the callouts and captions do), so designing graphics to carry a significant portion of the information makes good economic sense.

HOW TO DESIGN EFFECTIVE GRAPHICS

Writers, because they are comfortable with the world of words, may find planning and developing the text of a manual to be a relatively straightforward task, but may be somewhat baffled by how to manage the graphical elements. Too often, the graphics become an afterthought, added in as the page layouts permit. As a result, they may be disconnected from the information in the text, sized too small to see easily, or occur too infrequently to be of much help. For graphics to work well in a manual, they must be part of the manual design from the beginning, they must be appropriate for their function, and they must be designed to be readable.

Plan Text and Graphics Together

The most effective manuals are planned from the beginning as an information package containing both text and graphics. As you start to block out the initial outline for a manual, be thinking about what graphical elements you would like to include. Whenever you find yourself thinking "in pictures," plan to give the reader a picture as well.

Storyboards

A technique borrowed from filmmaking can be very useful in helping to integrate text and graphics in the early stages of designing a manual. Filmmakers prepare one or more "storyboards" for each scene of a film. These storyboards contain both visual and textual information. The technique translates easily to manual design: think of it as a step between the outline and a first draft. It works particularly well if you are planning a modular organization, as discussed in Chapter 4.

To make a storyboard, choose a small section of the manual — no more than a page or two in the finished manual — and prepare a plan for that section. The plan includes a prose summary of the material to be covered in that section and a sketch or description of an accompanying graphic. It's a good idea to jot down the *purpose*, as well as the subject, for both text and graphic. You can then lay out these storyboards in order, and see how the manual will develop. It is easy at this stage to rearrange sections if another order makes more sense. In any case, with storyboards, both writer and artist can be involved at the earliest stage of the manual, helping to ensure that text and illustrations will balance and support each other.

Functions of Graphics

We said that a storyboard should include a statement of the purpose of a graphic as well as its subject. The difference between the two is subtle, but real. The subject tells what the graphic is *about*, while the purpose is what the graphic is *for*. Graphics in manuals generally fall into one of three categories, depending on how they relate to the accompanying text. Regardless of the specific type of graphic (e.g., photograph, table, etc.), all graphics have one of three functions:

- To complement text
- To supplement text
- To substitute for text

Graphics that complement text are those that are necessary for the reader to understand what the text is saying. Figure 5.2 shows a page from a manual for a John Deere air seeder. Note how each instruction is accompanied by a photograph that directs the user how to carry out the instruction, by locating a lubrication point or inspection point. The instructions would be unintelligible without the photographs. Thus, in this situation, the graphic complements the text — which would certainly be incomplete without it.

This method of coupling a photograph with each specific instruction (called Illustruction™ by Deere) has been widely used and adapted in many industries (see, for example, Figure 5.3). It has the advantage of making the manual appear very accessible, even for poor readers. It also reduces translation costs, since some of the burden of communication is on the photograph rather than on the text. Because these text-and-graphic blocks are developed and stored electronically as a single unit, they make modular organization easier than if the text and graphics were developed separately.

Lubrication and Maintenance

4. Inspect bearing lock collars (A) (8 places) for tightness.

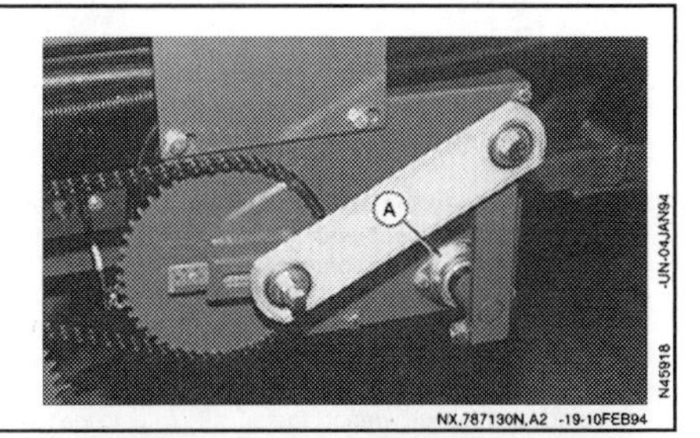

INSPECT AIR SYSTEM SEALS

Inspect air system seals (A) daily. Replace as necessary.

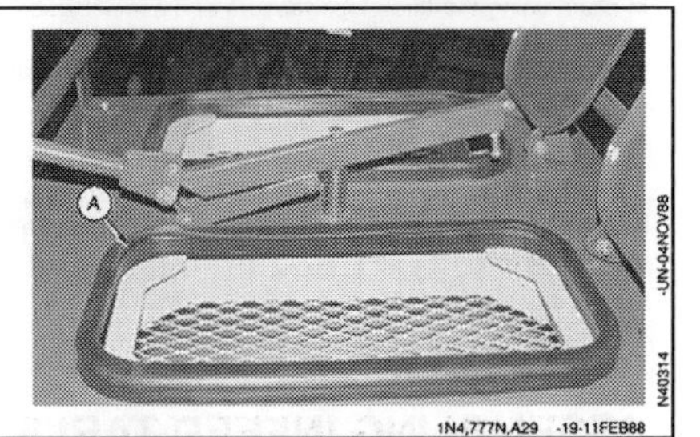

LUBRICATE DRIVE CHAIN

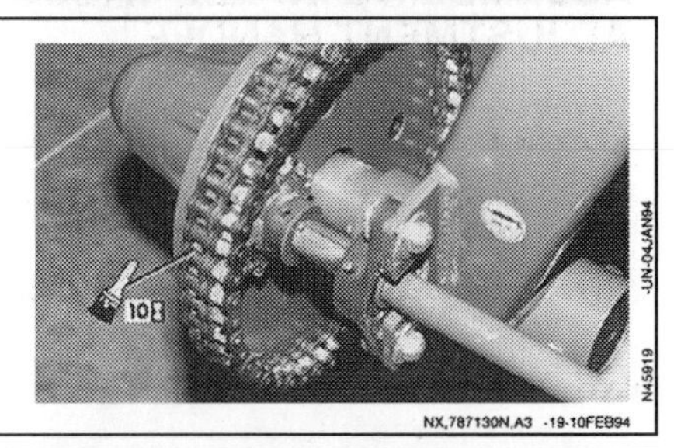

LUBRICATE METER DRIVE CHAIN

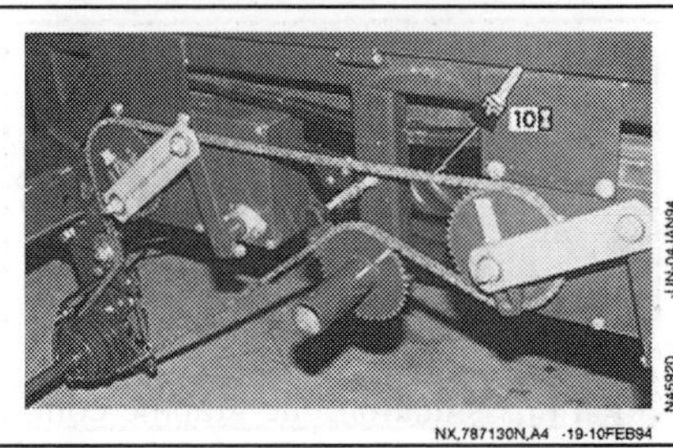

Figure 5.2 Use of graphics to complement text. Without the graphics to illustrate what the text is describing, the text would be useless. (From *Operator's Manual, 130 Bushel 787,* p. 40-5, Deere & Co., Moline, IL © 1994. All rights reserved. With permission.)

This format does have its limitations, however. Not all instructions can be broken down into small, simple steps. Sometimes complex explanations are necessary. And many kinds of instructions (using software, for example) do not lend themselves well to accompanying photos.

ASSEMBLING JOINTER TO STAND

1. **WARNING: JOINTER WEIGHT IS APPROXIMATELY 175 LBS. CARE MUST BE TAKEN WHEN LIFTING JOINTER ONTO STAND. A MINIMUM OF TWO PEOPLE WILL BE REQUIRED TO LIFT THE MACHINE.**

2. The infeed end of the jointer is fastened to the two holes (A) Fig. 9, and the outfeed end of the jointer is fastened to hole (B) on the two top end braces. **NOTE:** Dust chute (C) is on outfeed end of jointer. Line up the three threaded holes on the bottom of the jointer with the three holes (A) and (B) in the stand end braces.

3. Using the supplied wrench, fasten the jointer to the top of stand using the three lockwashers and special studs. Two of the special studs are shown at (D) Fig. 10, for the infeed end of the machine, and one special stud is shown at (D) Fig. 11, for the outfeed end of machine.

4. Once the jointer is completely secured to stand, push downward on the top of jointer until the stand adjusts to the floor surface. Then using the supplied wrench, tighten all stand hardware.

Fig. 9

Fig. 10

ASSEMBLING INFEED TABLE ADJUSTMENT HANDLE

1. Turn locknut (C) Fig. 12, clockwise on infeed table adjustment handle (B) as far as it will go.

2. Thread handle (B) Fig. 12, into block (D) which is located below infeed table (E).

3. Turn and tighten locknut (C) Fig. 13, against block (D).

Fig. 11

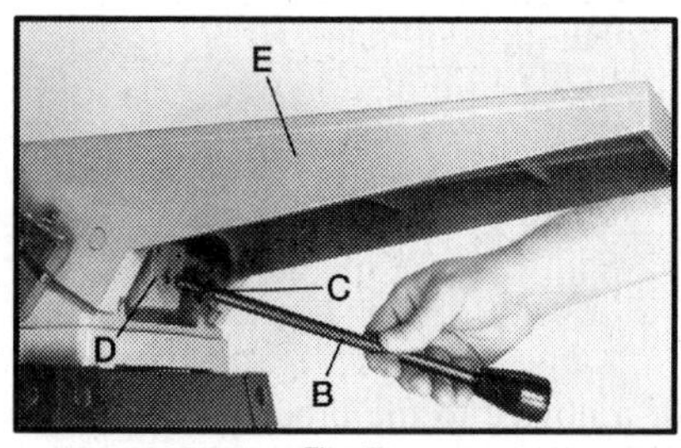

Fig. 12

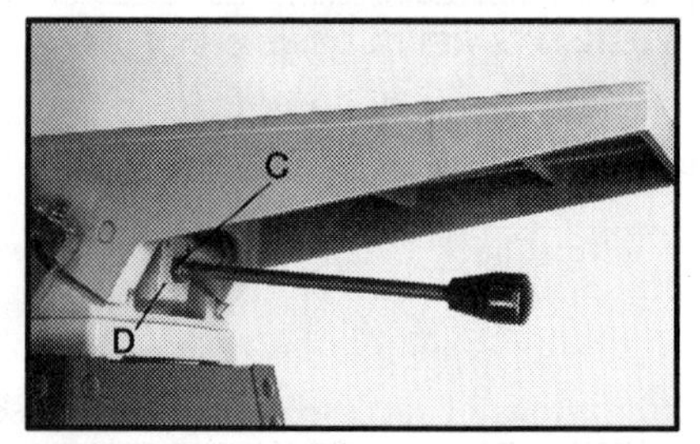

Fig. 13

Figure 5.3 Variation on the "illustruction"-type format. Here the idea of coupling photographs closely with the text is used, but with a different layout. (From *Instruction Manual, 6″ Deluxe Jointer (Model 37-190)*, Delta International Machinery Corp., Pittsburgh, PA, 1998, p. 9. With permission.)

Graphics that supplement text are those that serve primarily as illustrations: they add to the message presented in the text, but are not necessary for its understanding. A good example of such a graphic is shown in Figure 5.4. The photograph illustrates

PLACEMENT OF HANDS DURING FEEDING

At the start of the cut, the left hand holds the work firmly against the infeed table and fence, while the right hand pushes the work toward the knives. After the cut is underway, the new surface rests firmly on the outfeed table as shown in Fig. 64. The left hand should then be moved to the work on the outfeed table, at the same time maintaining flat contact with the fence. The right hand presses the work forward, and before the right hand reaches the cutterhead, it should be moved to the work on the outfeed table.

CAUTION: NEVER PASS HANDS DIRECTLY OVER THE CUTTERHEAD.

JOINTING AN EDGE

This is the most common operation for the jointer. Set the guide fence square with the table. Depth of cut should be the minimum required to obtain a straight edge. Hold the best face of the piece firmly against the fence throughout the feed as shown in Fig. 65. Maximum depth of cut should not be more than 1/8″ in one pass.

DO NOT PERFORM JOINTING OPERATIONS ON MATERIAL SHORTER THAN 10 INCHES, NARROWER THAN 3/4 INCH, OR LESS THAN 1/2 INCH THICK (REFER TO FIG. 66).

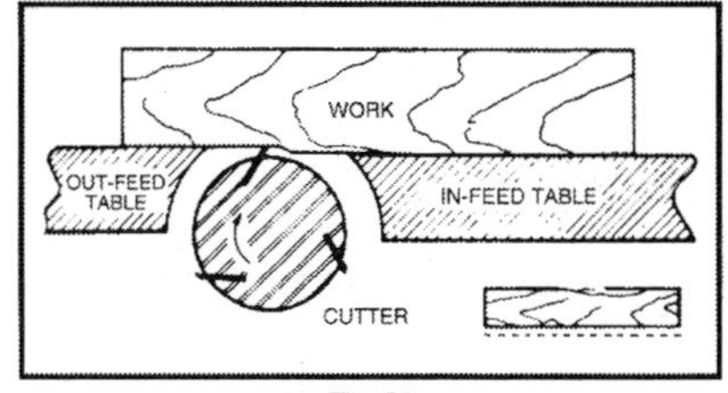

Fig. 64

Fig. 65

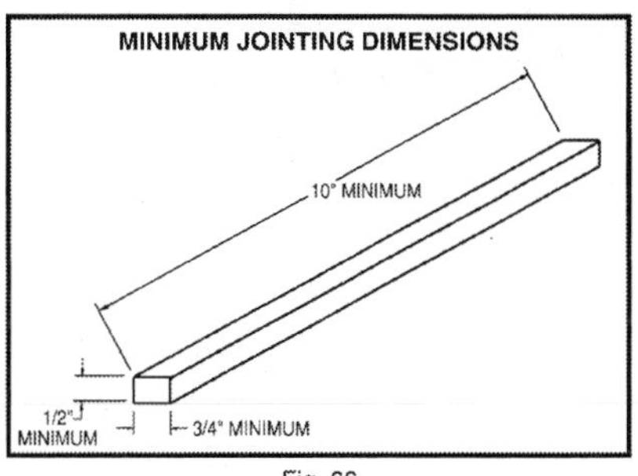

Fig. 66

Figure 5.4 Use of graphics to supplement text. In this example, the text is easily understood without the accompanying photographs, but they certainly serve to enhance it, particularly for a first-time user. (From *Instruction Manual, 6″ Deluxe Jointer (Model 37-190)*, Delta International Machinery Corp., Pittsburgh, PA, 1998, p. 22. With permission.)

what the text describes, but the text still makes sense without it. While graphics that supplement text may not be required for the text to be understood, they certainly contribute to it. Such graphics are especially useful for first-time users.

Remember that no reader, regardless of how experienced, is as familiar with the product as you are. Having written the manual and (if you are an engineer–writer) possibly having designed the product as well, it is easy to forget the questions and confusions of the user, especially those of a novice to the product. Try to look at your product with fresh eyes — imagine it is your first look — and think about what photographs and drawings would help you to understand the instructions. This is simply another form of writing with the user in mind. In any case, don't skimp. Use as many graphics as you can to supplement the text. The short-term added cost will save money in the long run by reducing the need for technical assistance and service calls. As we have noted elsewhere, an easy-to-use manual will help build repeat business, as well. Good illustrations are a big part of making a manual easy to use.

Graphics that substitute for text are becoming more common. While few manufacturers design manuals that are all pictures and no words (we have seen a few), many use graphics as a substitute for text in selected places. Figure 5.5, for example, uses an illustration to show how to hook up a scanner to a computer. The accompanying text describes USB connectors and how to configure the system to accept the new hardware, but does not instruct the user on the actual hookup.

When graphics substitute for words, they must be very carefully designed. If the graphics are unclear in any way — too small to identify parts, poorly reproduced, too cluttered — without text as a backup, the reader will be lost. It would be wise to test any such stand-alone graphics before using them in a manual. Be sure to pick a range of test subjects, and include some without prior experience with the product. See Chapter 6 for more information about usability testing.

Select the Appropriate Type of Graphic

A manual designer has a wide variety of graphics to choose from: photographs, line drawings of various kinds, graphs, charts, tables, and so forth. Each of these types of graphics has advantages and drawbacks. The trick is to choose the right kind of visual for the purpose you have in mind.

Photographs

Photographs are among the most common kinds of graphics used in manuals. Their chief advantage is that they are easily understood by the most technically naive user. A photograph of a product or part *looks like* that product or part. The user needs no special training to interpret a photograph. Additionally, photographs make

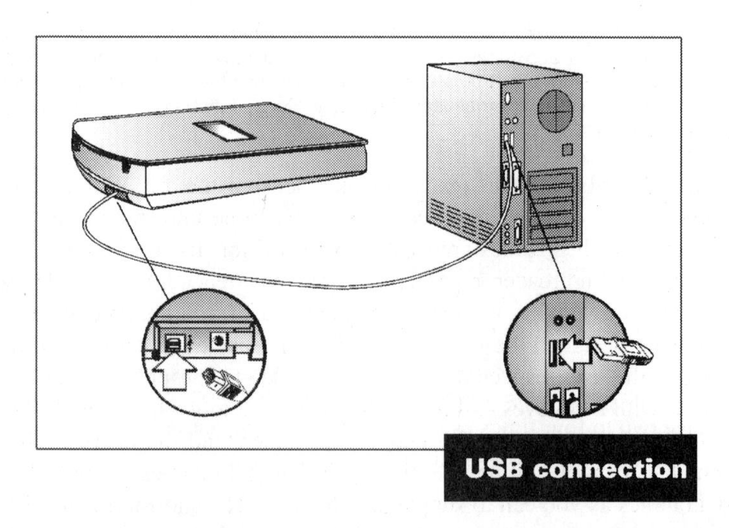

Figure 5.5 Use of graphics to substitute for text. In this example, showing how to hook up a scanner to a computer, the illustration stands on its own. No words are necessary to understand the connection procedure. (From *Installation Supplement, HP ScanJet Scanner*, Hewlett-Packard Company, Palo Alto, CA, 1998, p. 10. With permission.)

it easy to show a part in context, surrounded by other parts of the product. This can aid a new user in identifying unfamiliar parts.

The realism of photographs carries with it two drawbacks. First, photographs may easily be cluttered and fail to isolate the intended part from the context. Second, photographs do not lend themselves to hidden views. If you can actually saw a part in half, you could photograph the resulting "cutaway" view, but drawings are more commonly used to show hidden views.

Until recently, photographs had another disadvantage: they were expensive to reproduce, because they required an extra step before printing. Because photographs contain "continuous tones," that is, shades of gray, a process is needed to convert those grays to a form that can be printed with black ink on white paper. Photographs used to be projected through a *halftone screen*, which converted the image to a pattern of dots of variable size and spacing. At normal reading distance, our eyes would see shades of gray rather than individual dots.

If a photograph needed any retouching, such as airbrushing, that added to the cost of production as well.

As with so much else in the world of technical publications, the computer has changed all of this. While halftone screening is still available, it is much more common for companies to use a scanner to digitize a photograph — or bypass film altogether and make the shot with a digital camera that stores the image in electronic form. Today, using photoediting software, anyone can optimize exposure and retouch photographs. As a result of these innovations, it is usually cheaper to use photographs in a manual than to use drawings, which require the services of a graphic artist to produce.

Nevertheless, if you use photographs, you must deal with their weaknesses. As mentioned earlier, the realism of photographs can sometimes make it hard to isolate the important part from the surrounding context. One option may be simply to put a contrasting outline around the part or point it out with an arrow. Or you can airbrush (or retouch on the computer) the background distractions.

Whether you screen photographs in the traditional way or digitize them on a computer, you must be careful that they reproduce well in the actual manual. Even the best photograph can look fuzzy and out of focus if it is printed on porous paper that allows the ink to "bleed" or if it is shot through too coarse a screen. Film that is too "fast" will produce pictures that look grainy, a shortcoming that is only increased when the print is enlarged. Similarly, pictures shot with a low-end digital camera will also look grainy, but for a different reason. Here the issue is the resolution of the image, usually expressed as dots per inch, or dpi. When an image is digitized, it is divided into tiny dots. The more dots per inch, the finer the image. The cheaper digital cameras may only have 400 to 600 dpi, while the more expensive commercial ones may have two to four times as many. The lower-end cameras are fine for taking a picture of the new baby to e-mail to the relatives, but are not good enough for photographs in a manual.

Be sure that photographs used in a manual have adequate contrast and proper exposure. To some degree, problems in these areas can be fixed in the laboratory or on the computer, but they are best addressed initially when the shot is taken. Try to arrange for good lighting. Beware of shadows that obscure the very detail you are trying to show, or glaring highlights that ruin the light balance and make everything else too dark. Walk around the product to find the best angle to shoot from. Think about whether a single photograph will do, or whether you need a wide view to show something in context and a close-up for detail. As we suggest throughout this book, put yourself in the user's shoes. Plan your photographs to serve the user's needs.

Drawings

Unlike photographs, which are continuous-tone art (containing shades of gray), drawings are *line art*, meaning that they can easily be printed as a series of solid-color (usually black) ink lines on a page. Drawings do not require special treatment — camera-ready line art is no different from camera-ready text. Both can be turned directly into a printing plate with no special treatment. Ease of reproduction is only one advantage.

Drawings are widely used in manuals for a number of reasons. One big reason is that you can design a drawing to show exactly what you want, whether that view is normally visible or not. You can clear away extraneous clutter that is there in real life. You can show a cutaway view to point out hidden parts. You can create an assembly diagram, also called an "exploded view," to show how parts go together. See Figures 5.6 and 5.7. Drawings can be created relatively easily on a computer and stored electronically.

So why doesn't every manual use only drawings? The answer, of course, is that they have drawbacks as well as virtues. One major drawback is that they are expensive to produce, since they require a graphic artist to draw them. The artist may work with pen and ink or, more likely these days, mouse and monitor, but either way a professional will design and render a much better graphic than an amateur. The professional will also do it faster. Do not try to save money by having the writer (whose expertise is words) try to create something usable with a computer draw program. The writer will spend hours and the result is likely to be disappointing. This is not to say that the writer/artist with skills in both areas does not exist — just that such gems are rare.

To solve the cost problem, manual designers are sometimes tempted to use design drawings for the product as manual illustrations. After all, these already exist and they were drawn (or produced on a CAD system) by professionals. Except for certain very specific applications (such as in installation and maintenance manuals), design drawings are not suitable for manual illustrations. Any drawing is an abstraction and requires some visual sophistication to interpret. A photograph, we noted above, looks like the product. It is relatively easy for anyone to understand a photograph. Drawings, on the other hand, typically look quite different from the product. Many of the visual cues, such as shadows, foreshortening, and texture, that we rely on to give an image its "three dimensionality" may be absent from a drawing. They are almost certainly going to be absent from a design drawing. See Figure 5.8 for an example of a design, or engineering drawing.

In addition, design drawings are generally quite large, as big as 18 × 24 inches or even larger. To fit onto a typical manual page, they would need to be reduced drastically. When drawings are photoreduced, detail is lost, they become difficult to

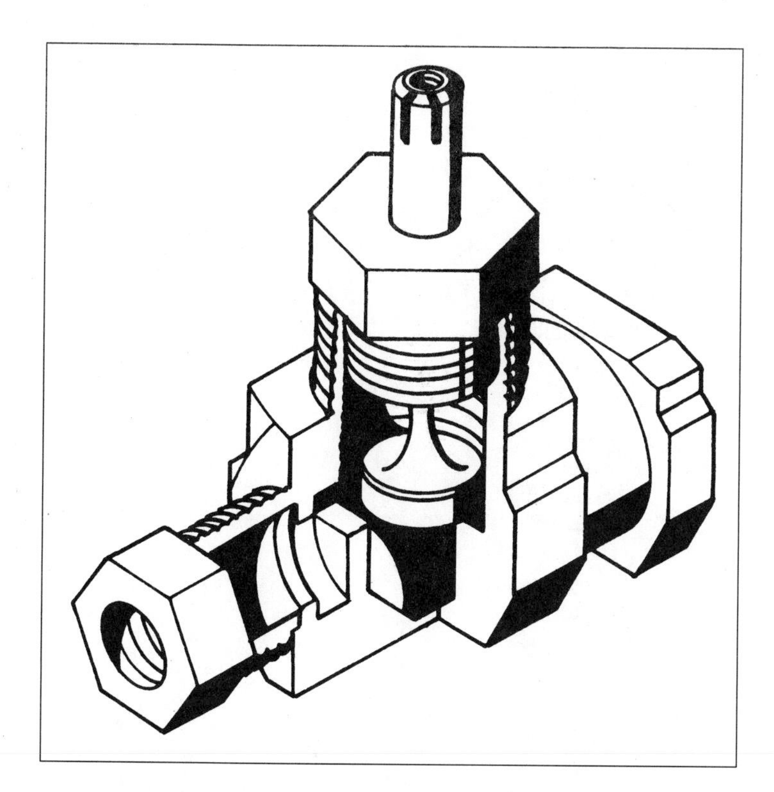

Figure 5.6 Cutaway view of a valve. A cutaway view allows the reader to see the construction on an object by presenting it as if a portion were literally cut away, revealing the layers of composition. The cutaway view is particularly useful to show the interior of something that is not normally disassembled. (Drawing by Teresa Sprecher.)

read, and the accompanying lettering becomes impossible to read. If you ever do use a design drawing as an illustration in a manual (an electrical schematic in a service manual, for example), consider redoing any callouts or other lettering *after* the reduction has been made.

The other option, used commonly in industrial maintenance manuals, is to keep the drawing its original size, and make it a foldout. This option should be used sparingly, and certainly not ordinarily in an operator manual. Odd-size pages are difficult to manage and easily get torn.

For many applications, the best choice of line drawings is a simple perspective drawing, as shown in Figure 5.9. This type of drawing adds back in some of those visual cues that make it easy to interpret, but it has the added advantage of allowing you to focus on the important part and reduce background clutter.

Charts

Charts differ from drawings in that they are symbolic rather than representational. In other words, charts use geometric shapes (circles, bars, diamonds, etc.) to stand for something rather than presenting a realistic image of the product. Many different

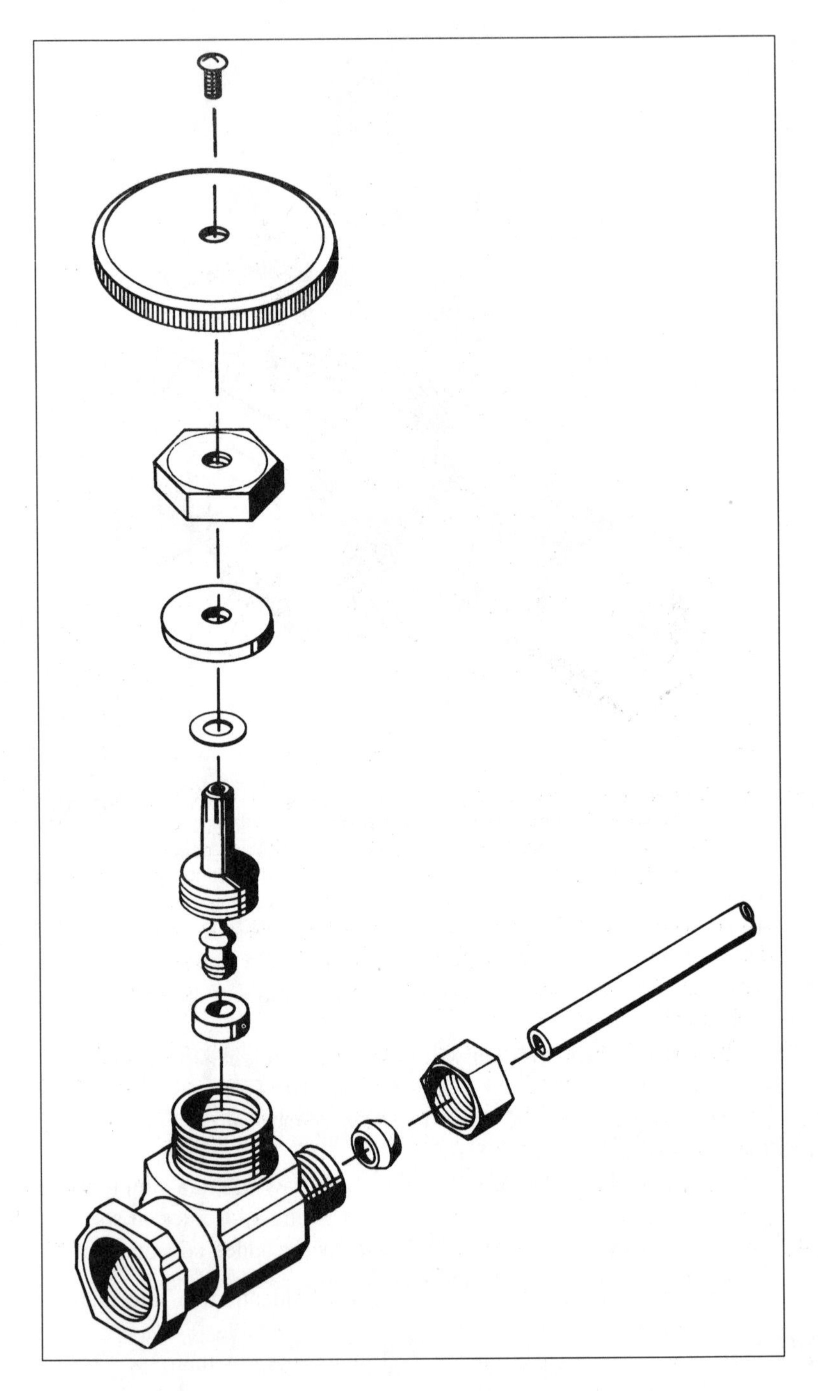

Figure 5.7 Assembly diagram of a valve. An assembly diagram uses an "exploded" view to let the reader see how the parts of an assembly fit together. It is used most often in conjunction with instructions for assembly or disassembly procedures. (Drawing by Teresa Sprecher.)

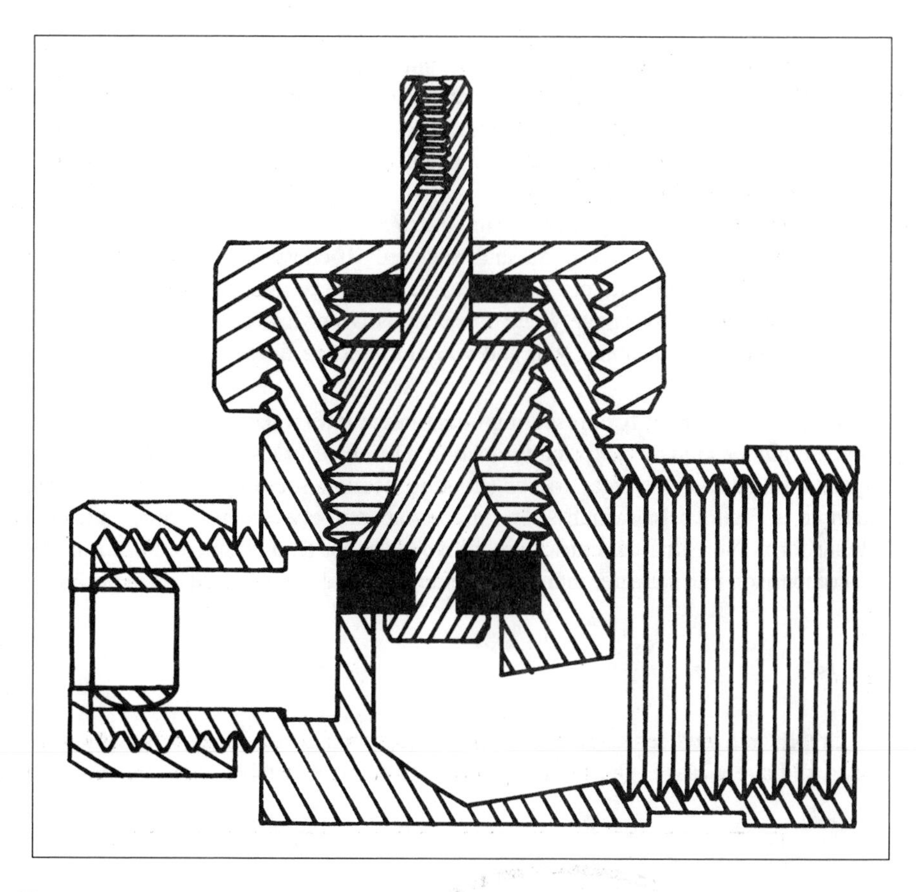

Figure 5.8 Design drawing of a valve. This design drawing, also called an engineering drawing, of the valve is much more abstract and requires special training to interpret. (Drawing by Teresa Sprecher.)

types of charts can be used in manuals, from the familiar bar chart showing, for example, the appropriate temperature ranges for various lubricants (see Figure 5.10) to a flowchart for a diagnostic procedure (see Figure 5.11). Whatever sort you decide to use, make sure it will be easy for your reader to understand. Keep it focused, uncluttered, and make sure that there is sufficient white space so that your reader can easily find the needed information.

Tables

Tables allow you to show a great deal of quantitative information in a very compact form. For example, Figure 5.12 shows a table that gives cooling system capacities for 11 different engines — all in a space less than 3 × 4 inches. Imagine the confusion that would result if you tried to present all that information in paragraph form. Useful as tables are, they must be designed well if they are to work well. The following list presents some guidelines for constructing tables:

- Arrange the headings and data in a rational order.
- Display items vertically that you want your reader to compare directly. Most people find it much more difficult to compare information that is arranged horizontally (see Figure 5.13).
- Include units of measurements in headings, rather than in each cell of the table — it will reduce clutter.
- Align columns of numbers on the decimal point.
- Use lines to divide columns and rows only if confusion is likely without them. Lines add clutter and tend to draw the eye along them rather than across them. Try to use white space as a divider instead. See Figure 5.14 for examples of tables with and without dividing lines.

Tables are essential for presenting quantitative information, but that is not their only use. Other kinds of information can be presented in table form. For example, Figure 5.15 shows a typical troubleshooting section set up as a table. Another example is Figure 5.16, which shows a comparison between two printer modes in tabular form. Note that the entries are all words rather than numbers. Still, it makes the comparison much clearer than if the writer had tried to write it all out in paragraph form.

Graphs

Graphs are used less commonly in manuals than other forms of graphics, but they still have a place. Quantitative information put in a table may sometimes be

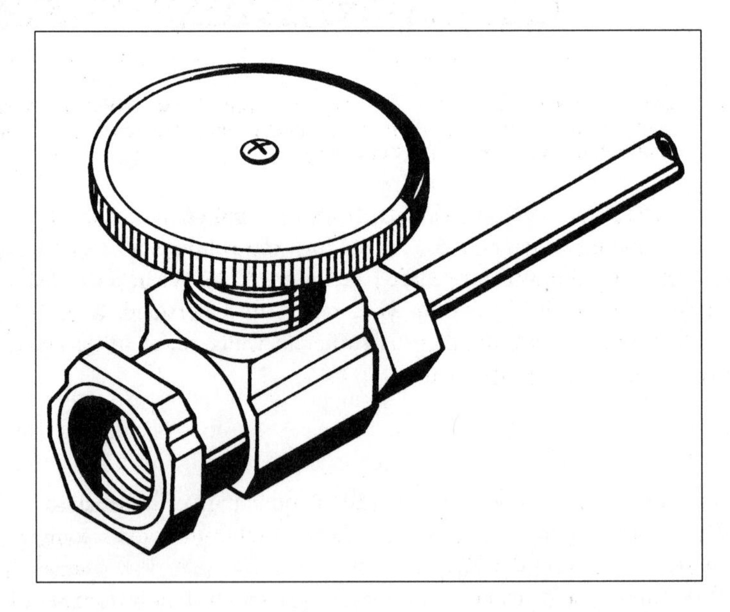

Figure 5.9 Perspective line drawing of a valve. This drawing is an abstraction — it is not *really* the valve — but it is quite recognizable to an untrained eye. (Drawing by Teresa Sprecher.)

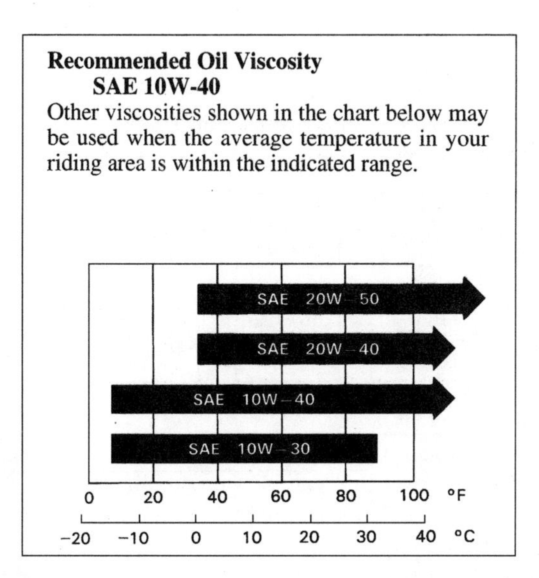

Figure 5.10 Bar chart used to show lubricants for different temperature ranges. (From *Honda Owner's Manual 97 VT1100C2 Shadow 1100*, American Honda Motor Company, Torrance, CA, 1996, p. 70, © Copyright by American Honda Motor Co., Inc. With permission.)

more effective if presented as a line graph. Consider, for example, the relationship between fuel consumption vs. speed for a car engine. You could show this relationship by creating a table, in which you listed various speeds in the first column and miles per gallon in the second. But because the relationship is not strictly linear — because of gear ratios, the engine may consume more fuel at the lowest speed than at a moderate one — the relationship would not be easy to sort out. By contrast, if you plotted the same information as a line graph, with speed along the x-axis and fuel consumption on the y-axis, the relationship becomes immediately obvious. Look at Figure 5.17, which shows these two options.

Design for Readability

Once you have decided on the purpose of a graphic, you can begin to make the design choices that will ensure the graphic works — because you know what is important. The essence of good visual design may be summed up in the following fundamental rule: *Make the important things stand out.* For example, you may have a choice between a photograph and a line drawing. As the previous section showed, the two media have different strengths. In one case, you may decide that it is most important to show the context of a part realistically, and choose a photograph. In another case, you may wish to show detail that would work better in a drawing.

As far as time and budget permit, make these decisions for each illustration individually. Don't use an old photograph from another manual just because it is handy. In the long run, an illustration or table designed with a specific purpose in mind will better serve the user's needs and the company's interests. You may find it useful to sketch (or ask the graphic artist to sketch) different versions of a drawing

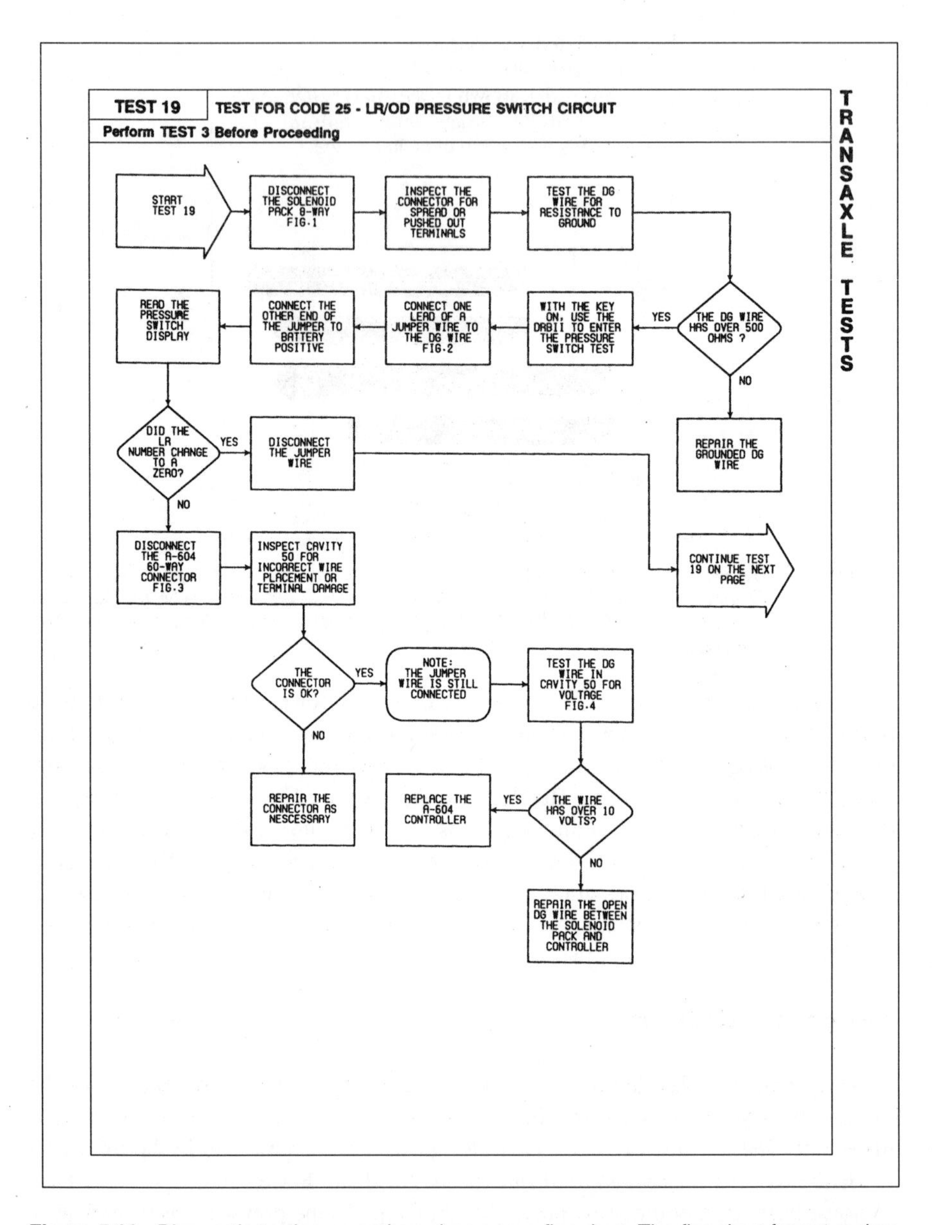

Figure 5.11 Diagnostic testing procedure shown as a flowchart. The flowchart format makes it very easy to see the sequence of tests. (From *Powertrain Diagnostic Procedures A-604 Ultradrive Automatic Transaxle (1989), No. 81-699-9009,* Chrysler Motors Corporation, Center Line, MI, 1988, 45. With permission.)

or angles for a photograph that you have in mind, so that you can pick the one that works the best. After all, you wouldn't expect to be able to sit down at the computer and write camera-ready copy on the first try — graphics need "rough drafts," too.

COOLING SYSTEM CAPACITY			
ENGINE MODEL	ENGINE	RADIATOR	TOTAL CAPACITY
8.2 Dash	12.5	19.0	33.5 quarts
6V-92TA Dash	24.5	26.0	53.0 quarts
6-71T Arrow	22.0	28.0	52.0 quarts
8V-71N Arrow	31.0	28.0	61.0 quarts
8V-71TA Arrow	32.0	28.0	62.0 quarts
6V-92TA Arrow	24.5	28.0	55.0 quarts
8V-92TA Arrow	29.0	28.0	59.0 quarts
NT 855	22.0	28.0	52.0 quarts
L10	13.5	28.0	44.0 quarts
3208 N&T	25.0	19.0	46.0 quarts
3406 N&T	24.0	28.0	54.0 quarts
NOTE: TOTAL CAPACITY INCLUDES APPROXIMATELY 2 QUARTS OF COOLANT NEEDED TO FILL HOSES.			

Figure 5.12 Table showing data in a compact form. The use of a table permits a great deal of information to be packed into a small space. (From *Pierce Chassis Operation Manual, Part Number 90-5000*, Pierce Manufacturing Co., Inc., Appleton, WI, 1990, 21.)

Visual design is a major area of study and certainly cannot be covered thoroughly in this book. The references at the end of this chapter give some suggestions for further reading in this area. The following general principles, however, apply to any visual presentation of information, regardless of type.

Make Them Big Enough to See

The user should easily be able to read and interpret a graphic at normal reading distance. Remember that a manual may be used in difficult conditions. Depending on the product, the user may be reading the manual in a basement or a dimly lit barn. Even if the lighting is good, it is crucial that the user be able to identify parts. If the graphic is too small (or excessively reduced to fit the page), a machine screw may be indistinguishable from a large bolt.

The ability to reduce and enlarge graphics is a wonderful thing, but it requires care in its use. In a photograph, excessive reduction will obliterate important detail. In a drawing, too much reduction can cause lettering to fill in, so that an *e* looks just like an *o*, and closely spaced lines to merge. Be particularly careful with large assembly drawings. Typically a line weight that works fine in the original becomes much too light when it is reduced to page size. If the original drawing was done on a CAD workstation, you can go in and change line weights to better suit the smaller size. If the drawing was done by hand, it would be better to redesign it specifically for the manual.

Often, you can solve the problem of loss of detail from reduction by adding a close-up or inset. For example, Figure 5.18 shows the use of a close-up to show

PARDEE SPINNING RODS

MODEL	LENGTH (FEET)	LINE WT (LB-TEST)	LURE WT (OZ)	ACTION	PRICE
UL-500	4.5	2-4	1/16-1/8	LIGHT	$32.50
UL-600	5	2-4	1/16-1/8	MEDIUM	$38.00
WB-650	6	6-8	1/8-1/4	MEDIUM	$43.00
WB-800	6.5	6-8	1/8-1/4	STIFF	$47.50
BN-850	6	8-12	1/4-3/8	MEDIUM	$52.75
MS-900	6	12-16	1/4-1/2	MEDIUM	$63.00
MS-1000	6.5	12-20	3/8-5/8	STIFF	$72.00

(VERSION A)

PARDEE SPINNING RODS

	UL-500	UL-600	WB-650	WB-800	BN-850	MS-900	MS-1000
LENGTH (FEET)	4.5	5	6	6.5	6	6	6.5
LINE WEIGHT (LB-TEST)	2-4	2-4	6-8	6-8	8-12	12-16	12-20
LURE WT. (OZ)	1/16-1/8	1/16-1/8	1/8-1/4	1/8-1/4	1/4-3/8	1/4-1/2	3/8-5/8
ACTION	LIGHT	MEDIUM	MEDIUM	STIFF	MEDIUM	MEDIUM	STIFF
PRICE	$32.50	$38.00	$43.00	$47.50	$52.75	63.00	$72.00

(VERSION B)

Figure 5.13 Two versions of a table showing the importance of vertical comparison.

detail. Be sure that if you use this technique, the user can tell where the part shown in the close-up view is located on the product. A nice close-up view of the power-steering fluid filler cap is not much use to car owners if they can't find the part when they look under the hood.

Give Each Graphic A Single Focus

Without question, the most common fault of illustrations in manuals is that they are cluttered. A user will find it difficult to focus on (or even figure out) what is important in a graphic overloaded with information. Illustrations and charts must be edited just like prose: figure out what the purpose of the graphic is and then include only what is necessary to fulfill that purpose. You should not include everything you know in a graphic — any more than you would in a paragraph. Simplify the visual

LINE CAPACITY OF SPOOL (IN YARDS)		
Lb. Test	Large	Small
2	---	---
4	400	300
6	350	200
8	300	150
10	250	100
12	200	---
15	150	---
18	100	---

LINE CAPACITY OF SPOOL (IN YARDS)		
Lb. Test	Large	Small
2	---	---
4	400	300
6	350	200
8	300	150
10	250	100
12	200	---
15	150	---
18	100	---

PARTS LIST (PARTIAL)			
Part Name	**Order No.**	**Part No.**	**Price**
Axle	305	81-214	$2.25
Baffle Plate	6342-R	56	1.20
Click Spring	11	322-D	0.30
Drive Gear	452-T	9120	4.50
Pivot	6783-DE	3	0.75
Rotating Head	66-2	81-615-A	11.45
Spool	43598-OS2	27	3.75
Transfer Gear	452 (453)	16 (17)	0.75 (0.85)
Trip Lever	340972	81-005	1.25

PARTS LIST (PARTIAL)			
Part Name	Order No.	Part No.	Price
Axle	305	81-214	$2.25
Baffle Plate	6342-R	56	1.20
Click Spring	11	322-D	0.30
Drive Gear	452-T	9120	4.50
Pivot	6783-DE	3	0.75
Rotating Head	66-2	81-615-A	11.45
Spool	43598-OS2	27	3.75
Transfer Gear	452 (453)	16 (17)	0.75 (0.85)
Trip Lever	340972	81-005	1.25

Figure 5.14 Two versions of two tables — with and without lines dividing the columns. Note that the use of lines is essential when the white space between columns is irregular — but a distraction when the white space forms a sufficient visual barrier.

PROBLEM	POSSIBLE CAUSE	POSSIBLE SOLUTION
No filling.	Filler nozzle is not primed.	Run about 4 tortillas through to prime.
	Filler metering cylinder is not primed.	Follow instructions under "Set-Up".
	Filler hopper interlock is not being made.	Make certain hopper is properly located & anti-rotation pin is fully engaged.
	Completion of stroke sensor has not been activated.	Use screw driver blade or wrench to push on pilot button.
	Air is off to machine or filler.	Check & turn on.
First fold & fill keep cycling.	Water droplets on electric eye lens.	Clean & dry lens.
	Residue buildup on 1st conveyor belt.	Clean off belt.
	Sensitivity set too high on 1st electric eye.	Adjust so eye only detects passing of tortillas.
Loosely folded burritos.	Tortillas are stale.	Use fresher tortillas. Best results are obtained when tortillas are only a few hours old.
	Back Stop on 2nd fold platen is too far back.	Adjust position to lessen surface area on platen.
	Too small amount of filling.	Adjust filler per pages 16 & 17.

Figure 5.15 Troubleshooting chart in traditional format. The Traditional format works well when there is no need for a sequential approach to troubleshooting. (From *Model 323-1 Tortilla Folder, Technical Service Manual,* Kartridg Pak Company, Davenport, IA, 1990, 34. With permission.)

presentation so that only essential items are included in detail and nonessential items are either absent or merely suggested.

If, despite your efforts to keep it simple, you still seem to have a complex illustration, consider splitting it in two — you may have more than one visual focal

	Epson GL/2	**HP LaserJet III**
PCL mode	Does not exist	Exists as the initial mode
Paper eject	Supports PG, AF commands	Supported in PCL
Auto eject	SelecType setting	Not available
Reduced printing	SelecType setting	Available in PCL
Switch to PCL (ESC % # A)	Not supported	Supported
Reset (ESC E)	Ejects paper and then initializes	Ejects paper, switches to PCL, and then initializes
PJL, EJL, and ES	Supported	Supported
Advance Full Page (PG, AF)	Supported	Not supported

Figure 5.16 Table used to make comparison clearer. This example, taken from a printer manual, compares two printer modes. Even though all the entries are words, the table makes the comparison much clearer than if the information were written in paragraph form. (From *Reference Guide, Epson ActionLaser 1000/1500*, Epson America, Inc., Torrance, CA, 1992, p. B-45. With permission.)

ENGINE EFFICIENCY

Miles Per Hour	Miles Per Gallon
10	30
20	37
30	38
40	37
50	33
60	27
70	20
80	10

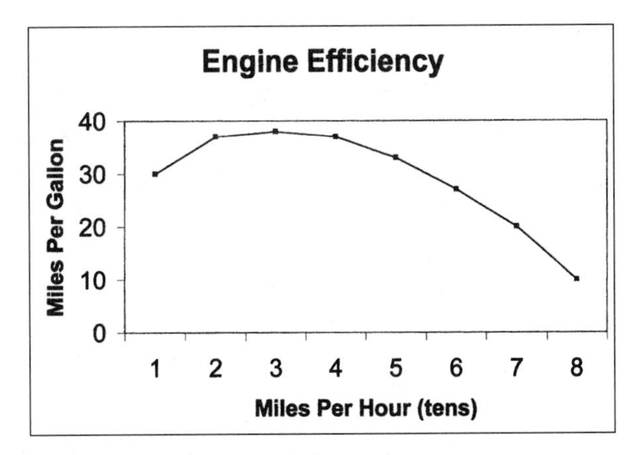

Figure 5.17 Numerical information shown in table and graph form. Notice how the relationship between the values is much easier to see when the data are presented as a line graph.

point. Break the presentation down by systems or show one overview and one or more close-ups. This is an especially important technique to use in parts catalogs, which too often show the entire product in a single incomprehensible assembly drawing.

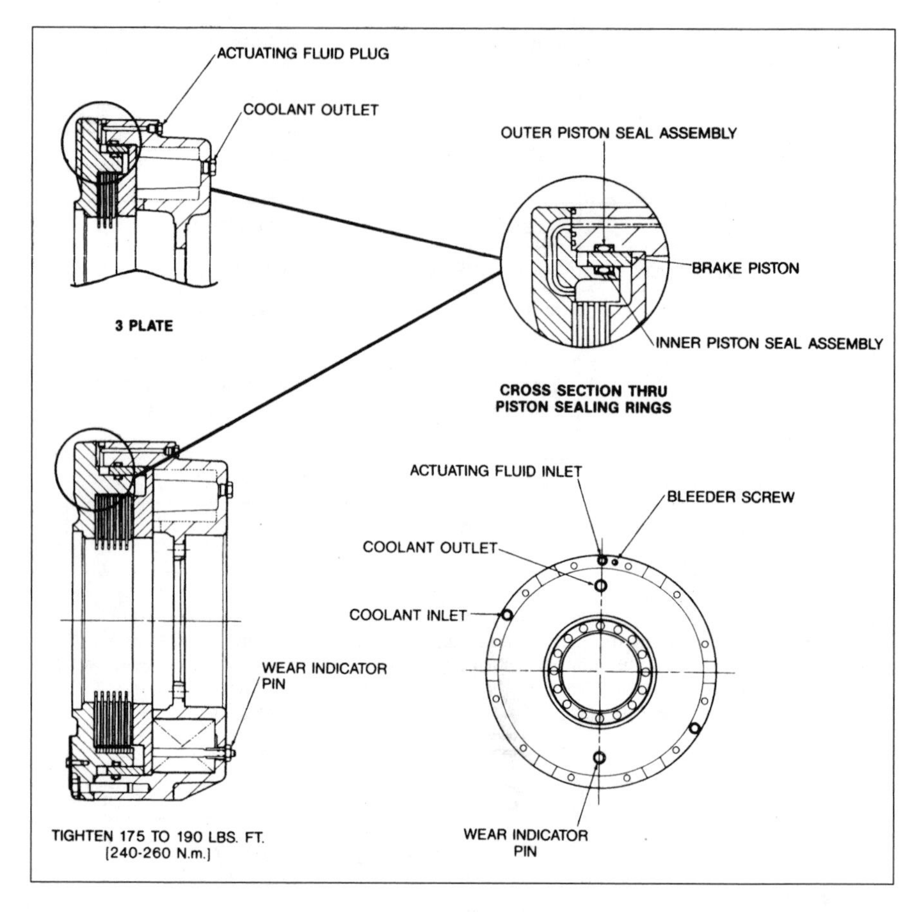

Figure 5.18 Use of a close-up to show detail. Note how the use of a close-up permits much more detail to be shown than would be possible otherwise. (From *LCB 13150/16150 Posi-Stop Liquid Cooled Brake Maintenance and Service Manual Supplement,* Clark Components International, Statesville, NC, 1989, 4. With permission.)

This process of simplification is terribly difficult — there is a great temptation to include more than you should — but you will find that knowing the purpose of the visual will help enormously. If you have a clear idea of what you want the drawing, photograph, or chart to accomplish, you can use that idea as a filter to screen out peripheral information, much as you would use a topic sentence to guide your choice of what to include in a particular paragraph. For example, the block diagram shown in Figure 5.19 is perfectly adequate for explaining the theory of operation of the transaxle in this service training manual. The full schematic (Figure 5.20) is included at the end for reference.

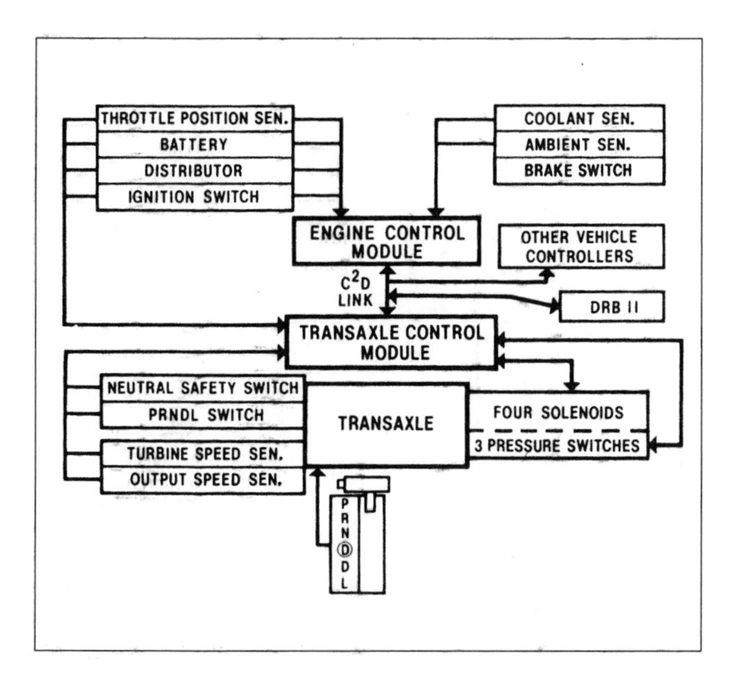

Figure 5.19 Block diagram to indicate electrical components. At this point in the manual, a block diagram is sufficient. Using a block diagram permits the reader to concentrate on the important parts and avoid being distracted by detail. (From *Powertrain Diagnostic Procedures A-604 Ultradrive Automatic Transaxle (1989), No. 81-699-9009,* Chrysler Motors Corporation, Center Line, MI, 1988, 19. With permission.)

Make Them Clear

Be sure that each drawing, photograph, table, and chart has a title that tells what it shows as well as a figure number or table number. For example, "Figure 3, Location of Idle Adjustment Screw" is much better than just "Figure 3." Pay particular attention to how you "call out" items in the illustration. Here, as always, there are trade-offs to be made. For example, compare Figures 5.21 and 5.22. They are very similar illustrations showing the controls for a motorcycle. The first one, from a Honda manual, uses part names and arrows to call out the parts; the second, from a Suzuki manual, uses numbers and lines (with no arrowheads) to call out the parts. The Honda illustration, because it uses part names, saves the user a step in getting the information. The user does not need to go to the legend to identify a part. On the other hand, the Suzuki illustration appears less cluttered and more focused on the drawing. Furthermore, it would require no change to the illustration itself for use in a translated manual — only the legend would need to be changed.

Which is better? The answer depends on your analysis of the audience and purpose for the graphic. Is it more important for the user to be able to look right at the drawing and identify parts without having to look away at a legend, or is it more

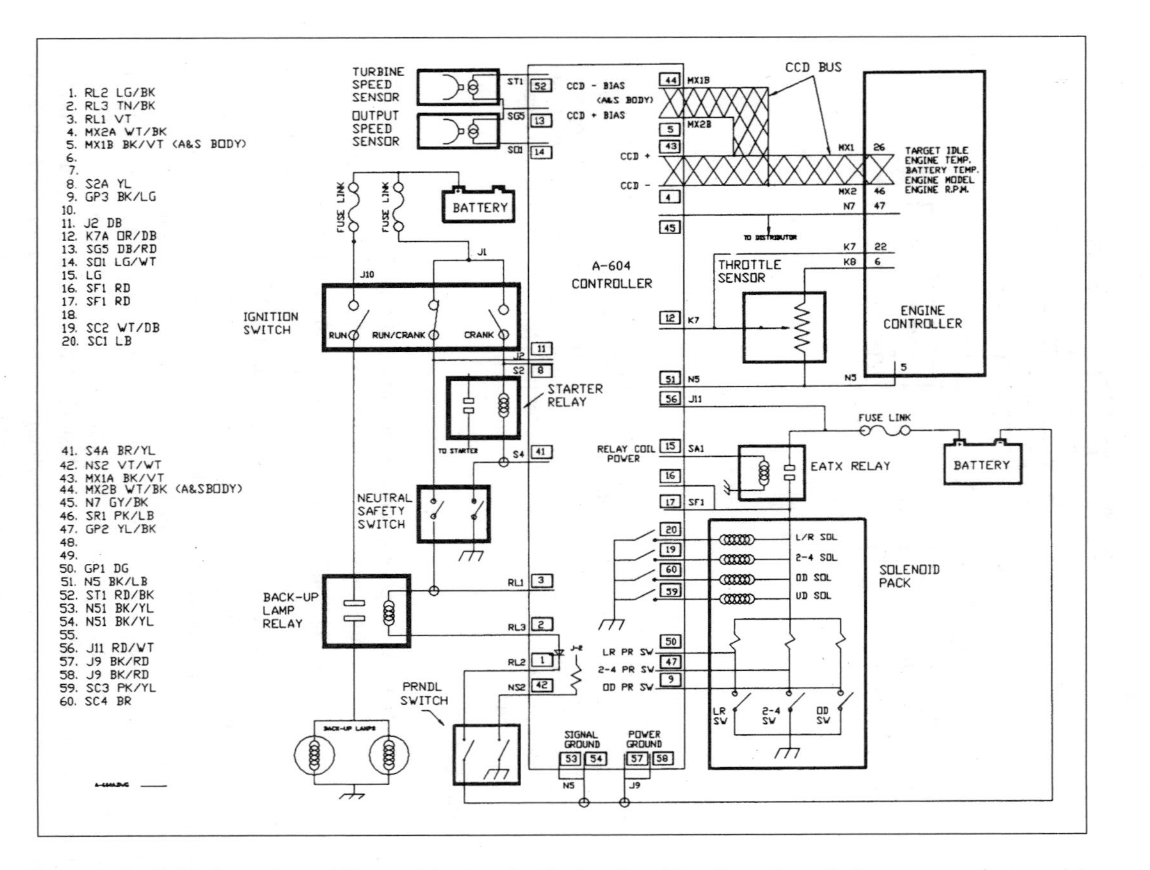

Figure 5.20 Full schematic to aid in servicing an electrical system. Here the schematic is necessary for completeness. (From *Powertrain Diagnostic Procedures A-604 Ultradrive Automatic Transaxle (1989), No. 81-699-9009,* Chrysler Motors Corporation, Center Line, MI, 1988, 118. With permission.)

important for the drawing to appear uncluttered? An experienced motorcycle rider might not be bothered by clutter, since he or she is already fairly familiar with the controls that are usually around the handlebars and would be looking primarily to find any changes. On the other hand, a novice might be holding the book in one hand, while finding parts on the motorcycle with the other. In that case, having to look at the legend would probably not be a problem, since the user would be looking back and forth between the product and the manual anyway. How important is it that the manual be easily translated? If marketing plans call for translation into several languages, the numerical callouts will save money.

Whichever way you choose, make it as easy for the reader as possible. Be certain that it is clear what parts the arrows point to. If you use numbers or letters to call out parts rather than words, arrange them in a rational order: try to put them in sequence. Usually you will have to choose whether the callouts are in sequence or the parts in the legend are in some order (such as alphabetical). Base your decision on how the graphic will be used. If the user will be looking at the illustration *first* to locate the part, as would be likely in a parts catalog or in a procedure description, make the numbers on the illustration sequential. On the other hand, if the user would first be looking up a name in the legend, and then locating it in an illustration, make the entries in the legend sequential (alphabetically or by part number). In either case, keep the legend close to the illustration — on the same page if possible, on a facing page if not. Never make the reader turn the page to look at the legend.

Use the principles of good visual design to help focus the reader's attention and to avoid confusion:

- The eye moves along lines, not across them. Use line direction to lead your reader's eye to the central focus of the graphic.
- Bigger or more detailed objects will seem more important than smaller or less detailed ones.
- Similar shapes (in a block diagram, for example) will suggest similar function.

Provide plenty of white space around and within a graphic. White space can help the user organize visual information, reduce clutter, and serve as a *de facto* "frame" around a graphic. Sometimes, all a confusing or cluttered illustration needs is a bit more white space for it to become more inviting to use.

Integrate Graphics and Text

At the beginning of this section, we recommended that you plan the graphics at the same time as you plan the text, using storyboards or some other similar system to ensure that the two modes of communication develop together. This alone will go a long way to ensuring that the graphics in the manual work well with the text. Graphics, like text, often convey information most effectively in a general-to-specific order. Just as you would begin a new section of text with an overview, you can provide an overall illustration of an assembly before you present detailed drawings of each component.

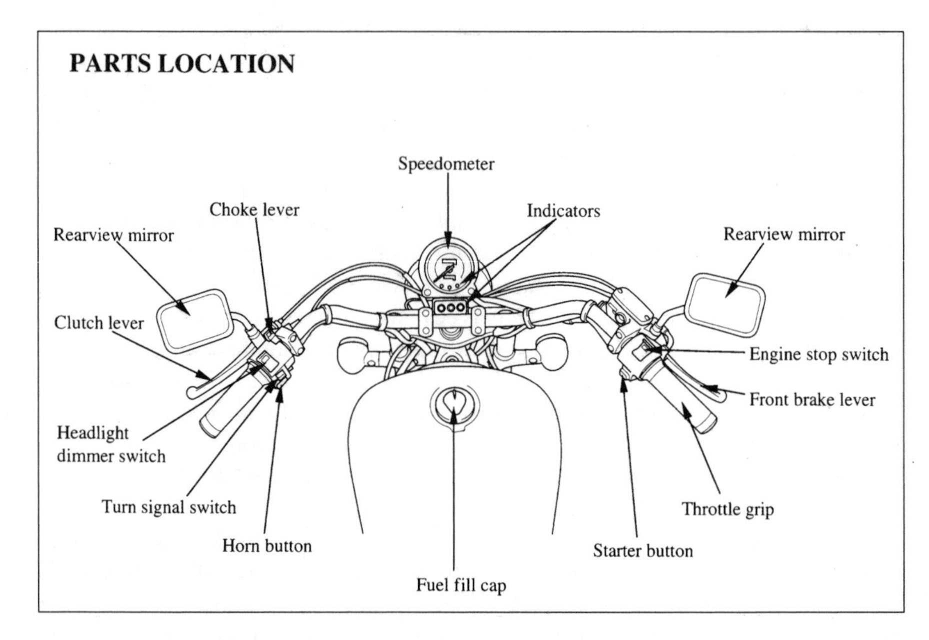

Figure 5.21 Illustration using words to call out part names. Using words and arrows to call out parts makes it quicker for the user to find the part name, but limits the amount of information that can be shown in one drawing and requires an extra step for translation. (From *Honda Owner's Manual 97 VT1100C2 Shadow 1100*, American Honda Motor Company, Torrance, CA, 1996, p. 11, © by American Honda Motor Co., Inc. With permission.)

When you begin to lay out the pages, you will appreciate having planned text and graphics together. In Chapter 4, we cautioned about allowing graphics to "leak" into the gutter between columns of text. In your initial planning, as you start to choose the kinds of graphics you will use, you may decide that a 2/5 format would serve your purposes better than a two-column format, for example. It is a great deal easier to change format at the beginning of a project than at the end.

In any case, be sure that you tie the text and graphics together:

- Always refer to the graphic in the text.
- Place the graphic as soon as possible *after* the first reference to it.

It is quite disconcerting to come upon a figure in text without explanation (particularly if it has no caption other than a figure number). Conversely, it is irritating to read "see Figure 24" in the text, and then thumb through page after page of the manual without finding it. Naturally, the "perfect" layout is seldom possible. As with so much else in manual production, laying out graphics and text requires balancing the ideal of perfection against the reality of time and budget constraints.

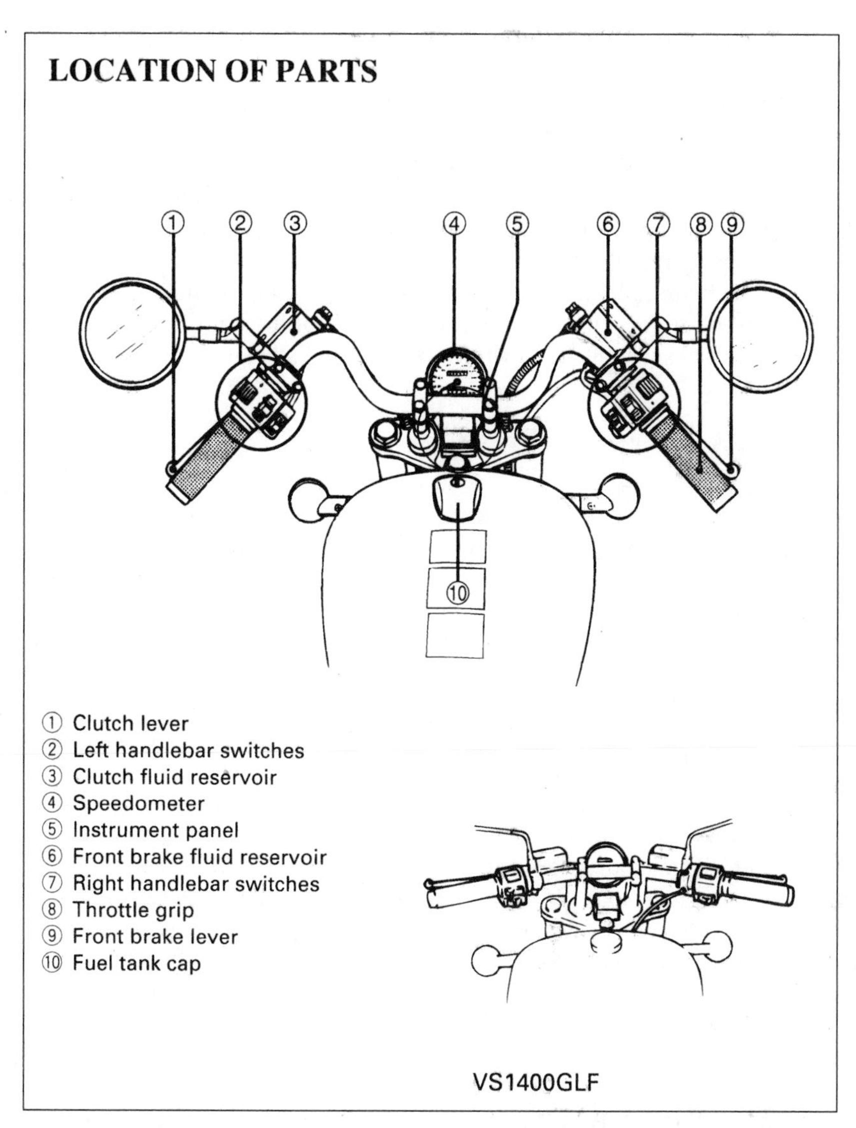

Figure 5.22 Illustration using numbers to call out part names. Using numbers to call out parts requires the reader to go through an extra step, but allows more information to be included in a single illustration and simplifies translation. (From *Owner's Manual, Suzuki VS1400GL*, American Suzuki Motor Corporation, Brea, CA, 1997, p. 10. With permission.)

NEW USES OF VISUAL MEDIA

Perhaps no area of communication has been more radically changed by the advances of computer technology than that of graphics. Although the computer has made text production much easier, the text itself hasn't really changed all that much: we still use typefaces invented in the 19th century (or earlier) and produce columns of text that look quite similar to those of a century ago. Graphics, on the other hand, have undergone a transformation.

Videotapes are widely used in industry as training aids in a variety of areas. Instructional videos that demonstrate how to set up or operate products sometimes accompany the manual. The laptop computer has made CD-ROM a practical alternative to bulky manuals for storing reference material such as parts catalogs. The development of the Internet has allowed manufacturers to put some of their documentation online — even for noncomputer products. All three of these media have their uses, but none is a total substitute for a written manual, and all require expertise to design well.

Videotapes

The obvious advantage of a videotape is that the viewer can actually watch a procedure being performed by an expert. It can be the next best thing to the "expert-in-residence" that we discussed at the beginning of this chapter. A viewer of a videotape does not need to be literate or skilled in interpreting technical drawings.

On the other hand, the videotape requires that the user have the appropriate equipment to view it. Further, unlike a paper manual, the user cannot easily "leaf through" a tape. For these reasons, we find that videotapes are most useful as training tools for service technicians or as supplements for one-time use, as in showing how to set up a product for operation.

To work well, videos must be carefully planned and produced — and these are not simple tasks. Developing a good videotape means putting a great deal of time into designing it and finding skilled technicians to shoot and edit the tape. Videotapes produced by amateurs look like home movies and almost always present information much too fast. If your company is thinking of using videos, it is well worth the investment to contract with professionals unless you already have the skills in-house. Do-it-yourself videos are usually much less effective than a well-written manual.

CD-ROM

A compact disk can hold enormous quantities of information in a very compact form. An entire encyclopedia, for example, can fit on one CD-ROM. CDs are tremendous space savers, especially for people who have to travel. A service representative, for example, who travels to various locations to provide service for industrial machinery, may need a good deal of reference information when out in the field. It is much easier to carry a laptop and one or two CDs than to lug along a whole library of reference books. When prices change, or models are redesigned, it's easier to send the customer a replacement CD than to revise and reprint a manual.

The major drawback, of course, is that, like videos, the user must have the appropriate equipment to use the CD. A laptop with a dead battery is useless unless there is a handy source of electricity nearby. CD-ROMs are read-only media, and that makes it difficult to make notations, as you might do in the margins of a standard book. Newer CDs allow the user to write on them as well, but these are not yet in wide use.

Perhaps the biggest obstacle to widespread use of CDs to substitute for paper manuals is the conditions under which manuals may need to be used. The dirty, dusty environment of a gravel separator would be very hard on a laptop computer. The tractor that needs to be fixed may be sitting in bright sunlight or in a dim machine shed — conditions that would make it difficult to see a computer screen.

Web Pages

In just a few years, the Internet has grown from an arcane idea of a handful of university scholars to an everyday presence for millions of people. Increasingly, manufacturers are using the Internet as a marketing tool, and some are relying on it for documentation for products. Online documentation can take advantage of hypertext — the ability of the user to mouse-click a word and instantly be given a screenful of linked, related information. Documents can, in essence, be arranged in whatever order the reader prefers. In addition, the reader can be taken, by hyperlink, not only to other pages, but to entire other documents for information. The possibilities are intriguing.

Updating online information is simple: the manufacturer just changes what's on the server from its end, knowing that users will get the latest version each time they use the site.

To design an online document well takes considerable expertise. Entire books have been written about this issue alone. As with indexing, the best online documents are designed to make it easy for users to get around and find the information they need. Design takes skill and the ability to conceptualize extremely complex patterns of association.

Online documentation suffers from the same limitations as CD-ROM. The user has to have the proper equipment and an Internet connection to use it, which may make it less portable than a manual. After all, you can prop a book up on one corner of a machine while you work on a nearby area. If it slips off, you just pick it up, dust it off, and find your place again. A laptop computer is not quite so durable. Advances in technology may make online documentation practical for a wide variety of products, but, so far, it lends itself primarily to software applications and related "desktop" products.

SUMMARY

As with text production, advances in technology have made it easier for manual designers to produce high-quality graphics for their manuals. Because graphics are so important in manuals — both to communicate certain kinds of information and to reach users that find text problematic — good planning and design are essential. Good planning means including graphics at the earliest stages of manual design, so that the graphics and the text function together as an integrated unit.

Good design requires choosing the best type of graphic for the purpose at hand and making sure that the final product is readable and clear. Different kinds of graphics have different strengths: a photograph may be better than a drawing if realism is important, but a drawing may work better to highlight a detail. Whatever type is used, it must be designed as carefully as prose to show just the right information in a manner that is clear and easy to read.

We ended the chapter with a discussion of an emerging trend: uses of nontraditional visual media such as videotapes and Web pages to supplement (or in some cases, substitute for) manuals. These high-tech alternatives present some exciting possibilities, but it is unlikely that they will ever entirely replace the traditional paper manual for most products. Even if they do, skill in visual design will remain an important and developing area for document designers.

CHECKLIST

- □ Have I looked at my product with a "fresh eye" — as if I were a novice user — and identified where to use graphics to help convey information?
- □ Have I planned my graphics and text together?
- □ Have I used the best type of graphic (photograph, drawing, chart, table) for what I want to show?
- □ Are the photographs shot from the best angle? Retouched as needed?
- □ Are the drawings simple and uncluttered? Appropriate to the audience?
- □ In tables, are the data arranged vertically for easy comparison?
- □ Have I eliminated unnecessary lines from tables to reduce clutter?
- □ Are the graphics big enough to see easily?
- □ If I have reduced drawings, is any text still readable?
- □ Have I called out parts clearly? Is what each callout refers to obvious?
- □ Are the graphics properly referenced in the text and placed close to the relevant text?
- □ Do the graphics convey an impression of concern for the customer, by being clear, inviting, and easy to use?

FURTHER READING

Bethune, James D. *Technical Illustration*. John Wiley & Sons, New York, 1983.

Cornsweet, Tom. *Visual Perception*. Academic Press, New York, 1970.

Jastrzebski, Zbigniew. *Scientific Illustration*. Prentice-Hall, Englewood Cliffs, NJ, 1985.

Richardson, Graham T. *Illustrations*. Humana Press, Clifton, NJ, 1985.

Tufte, Edward R. *The Visual Display of Quantitative Information*. Graphics Press, Cheshire, CT, 1983.

Tufte, Edward R. *Visual Explanations*. Graphics Press, Cheshire, CT, 1997.

CHAPTER 6

Safety Warnings

OVERVIEW

Few topics will generate more discussion, confusion, opinions, and frustration among manufacturers than that of safety warnings. We have all heard stories of people who received multimillion dollar product liability awards after suffering injury doing something patently stupid — like picking up a rotary lawn mower and trying to use it as a hedge trimmer, or attempting to dry a wet poodle in a microwave oven. As we talk with technical writers and editors, we find them both anxious for information on how to write good warnings and defensive about the need to do so: "You can't idiot-proof the product" is a common sentiment. The good news is that you don't have to. The cases alluded to above, as far as we can tell, *never happened.* We have been unable to locate a citation for either case. The first reference to them that we have found was in an insurance-industry advertisement explaining rate hikes. If any reader can locate a case citation for either of these, we would be delighted to learn of it.

And while there are, from time to time, multimillion dollar court awards, the reality is much more prosaic. According to the Consumer Federation of America, in the decade from 1987 to 1997 approximately 2.8 million persons filed product liability claims. Of these, just under half received no payment at all. The average award was $12,000 (inflated by the occasional multimillion dollar award) (Moll).

Nevertheless, product liability as it relates to instructions and warnings is a continuing concern for manufacturers, as it should be. Typically the development of safety messages, whether for inclusion in the manual or for labels on the product, falls to the technical writer. (One of the positive by-products of this concern with liability has been an overall improvement in the status of the technical writer and in the quality expected in manuals.) The technical writer, therefore, needs to be familiar with strategies for designing warnings that meet current standards and that communicate effectively to the user. In this chapter we provide a brief background on product liability law and the duty to warn. We offer guidelines for developing effective safety warnings for use in the manual and on the product. Finally, we address nonstandard formats for warnings, such as safety cartoons, and other safeguards for the manufacturer.

Remember that the warnings are just one part of a comprehensive effort to educate the user: other elements include marketing literature, the accompanying instructions, and technical support. The goal must be to teach the user to use the product properly and safely. If that effort is successful, and the product is not unreasonably dangerous, there will be no injury and no liability.

PRODUCT LIABILITY BACKGROUND

Product liability is a rapidly evolving area of law. New legal theories are applied to individual sets of facts and the courts duly accept or reject a particular interpretation. Sometimes the courts, particularly in different states, disagree with each other. Until a case makes it up the ladder to a higher court, that disagreement stands, so that a manufacturer may be liable in one state for actions that would produce no liability in another. Is it any wonder that writers of manuals are frustrated and confused? The area of product liability is extremely complex and changing, and the material in this chapter is necessarily general. As you write a manual, you should work closely with your company's product safety division and legal counsel to ensure that your part in the manufacturing process meets legal requirements.

Legal Theories in Product Liability

Product liability litigation may be based on one of three legal concepts: *negligence, breach of warranty,* and *strict liability in tort.* Negligence means that the manufacturer did not exercise reasonable care in the manufacture or marketing of a product, resulting in an unreasonable risk to the user. Breach of warranty means that the product did not do what the manufacturer said it would — but note that the warranty can be either express or implied. An implied warranty may be present in instructions for the product's use, even if the matter in question is not included in the written (or express) warranty. Strict liability in tort means that if a product is defective in manufacture or marketing, the manufacturer may be liable for damages even if the manufacturer was *not* negligent, was unaware of the defect, and made no claims about the product's performance. Strict liability concentrates on the product, not on the care with which the manufacturer operated, and it is the predominant legal theory applied in cases involving warnings.

At first, this concept seems unfair. How is it that a manufacturer or distributor can be held liable when it did nothing "wrong"? The rationale for strict liability is twofold. First, it is seen as an incentive for manufacturers to design safe products with adequate instructions and warnings. Second, it is seen as a way of sharing the cost both of injury and of safety. Because strict liability focuses on the product, not the manufacturer, it encourages manufacturers and sellers to insure themselves against liability claims. This in turn increases the cost of the product, and those increased costs are passed along to the buyer in the form of higher prices. Thus, a single manufacturer or consumer does not bear the whole cost of an injury caused by an unsafe product, and manufacturers and consumers who benefit from safer products also pay part of the cost of increased safety.

When Is the Manufacturer Liable?

Manufacturers (and sellers) have a duty to provide reasonably safe products. Products that are unreasonably dangerous are considered defective. Regardless of the nature of the defect, four things must be true for a product liability suit to be successful.

1. The product must have a defect.
2. The defect must be present when the product leaves the control of the manufacturer or seller.
3. Injury or damage must be incurred.
4. The injury or damage must have been caused by the defect.

All four must be true. For example, an electric drill that overheats dangerously with normal use is obviously defective, but there are no grounds for a liability suit unless someone is injured. If someone inadvertently knocks the defective drill off a workbench so that it falls and breaks his or her toe, there is still no basis for a claim because the injury was not caused by the defect.

A product defect may be a design defect, a manufacturing defect, a packaging defect, a marketing defect, etc. In this chapter we will deal only with defects involving instructions and warnings. According to the *Restatement Third, Torts: Products Liability,* instructions and warnings are inadequate and the product is defective when "the foreseeable risks of harm posed by the product could have been reduced or avoided by the provision of reasonable instructions or warnings … and the omission of the instructions or warnings renders the product not reasonably safe" (Restatement Third, §2.c).

Note that the standard is *reasonableness.* The manufacturer is not expected to warn of every conceivable hazard. What is reasonable? A professor who taught at the University of Wisconsin Law School was fond of answering that question by saying, "Go to court, and the judge will tell you." Even the authors of the Restatement Third acknowledge the difficulty: "No easy guideline exists for courts to adopt in assessing the adequacy of product warnings and instructions. In making their assessments, courts must focus on various factors, such as content and comprehensibility, intensity of expression, and the characteristics of expected user groups" (Restatement Third, §2. Comment i). Nevertheless, some general guidelines do emerge from the case law, and this chapter will present those. First, however, we want to address the purpose of instructions and warnings.

Two Approaches: Protect the Manufacturer or Protect the User

Throughout this book, we have emphasized that form follows function — the purpose of a document determines, in part, what it looks like. Sometimes distinctions between purposes are subtle. A colleague of ours is fond of telling the following story. During the Dark Ages in Europe, when medicine was not very advanced, it was not easy to tell the difference between death and deep coma. As a consequence, the populace had a considerable (and rational) fear of accidentally being buried alive. This problem was addressed differently in England and in Spain. In England, coffin makers developed elaborate signaling systems, so that if an unfortunate happened to wake up while being borne off to the cemetery, he could pull a string placed in his hand inside the coffin, and a bell would sound, alerting the pallbearers to the error. In Spain, they simply drove a stake through the heart of the person before placing her in the coffin. The two practices had subtly different purposes. In England, they made sure no one who was alive was buried. In Spain, they made sure that everyone who was buried was dead.

A similar subtle distinction exists in how manufacturers address the issue of warnings. One approach is to write warnings to protect the company from liability claims. These tend to read like the fine print in insurance policies. Examples 6.1 and 6.2 show warnings designed to protect the company. These use complex language and try to shift the burden from the company to the user. Contrast them to Example 6.3, which is designed to protect the user. Here the wording is straightforward, and helpful information is included.

We strongly recommend that you write your warnings with the goal in mind to protect the user. We think there are three very good reasons for doing so:

1. If you are successful in protecting the user, there will be no injury and no liability.
2. If your warning is unsuccessful, and the user is injured, a warning designed to protect the user will look a lot better to a jury than one designed to protect the company.
3. The current state of the law is that the manufacturer or seller *is* liable for defective products, whether or not it disclaims that responsibility, so the disclaimer-style "warning" won't help.

Example 6.1 Warning designed to protect the company.

> Provision of point-of-operation safety devices consistent with the use and operation of this machine is the sole responsibility of the user of this machine.

Example 6.2 Warning designed to protect the company.

> All persons authorized to use the equipment must be cognizant of the danger of excessive exposure to x-radiation, and the equipment is sold with the understanding that the ____________ company, its agents, and representatives have no responsibility for injury or damage which may result from exposure to x-radiation.

Example 6.3 Warning designed to protect the user.

> **WARNING:** This product contains chlorine. Mixing this product with other household cleansers, such as toilet-bowl cleaners, rust removers, and products containing ammonia, may release deadly chlorine gas. Do not use this product with other chemicals.

THE DUTY TO WARN

Manufacturers, distributors, and other sellers throughout the chain of distribution have a responsibility to provide safe products. Clearly, the best choice is to design hazards out of the product. If the design can be changed to eliminate, or substantially reduce, the hazard, there will be fewer injuries and less need for warnings. An example of a design change to reduce a hazard is the so-called "dead-man" switch on power lawn mowers. Any lawn mower manufactured for sale in the United States must be designed so that if the operator lets go of the handle, the blade stops within 3 seconds. Rather than warning against clearing debris clogging the discharge while the mower is running, the dead-man switch makes the hazardous act difficult to do.

Sometimes hazards cannot be designed out. An example might be a printing press, in which the turning rollers produce ingoing nip points, where an operator's hand could be drawn in and crushed. Designing out that nip point would mean that the press could not do what it was intended to do. In that case, an alternative is to provide a guard that shields the nip point while the press is operating. Another example might be a meat-slicing machine, in which an electrical interlock prevents the operator from opening the housing around the blade while the blade is in operation.

If a hazard cannot be designed out or guarded against, the manufacturer must provide a warning. But the courts have held repeatedly that a warning is not a substitute for good design or adequate guarding. For example, in the case of the meat slicer discussed above, it would not have been acceptable for the manufacturer to leave the blade unguarded, but to provide a warning — even a "perfect" warning.

Who Must Be Warned?

The manufacturer has a duty to warn anyone who might reasonably come into contact with the product. Thus, the manufacturer may be held liable for injury to someone other than the person who actually bought the product. For example, the buyer of industrial chemicals may be a manager or purchasing agent. The actual users will be workers on the floor. The buyer of a fleet of golf carts may be the owner or manager of a golf course — but the users will be golfing customers. This requirement that warnings reach the actual user of a product may mean that warnings need to be located on the product itself as well as in accompanying literature.

The situation for the technical writer is further complicated by the fact that the potential users of a product may comprise a very diverse group. You must consider the possible range in terms of age, sex, expertise, familiarity with product, even literacy. In *Hubbard-Hall Chemical Co.* v. *Silverman*, the court held that a written warning was not adequate because it failed to provide for illiterate users.

More commonly, the issue is not illiteracy but inability to read English. In the United States, the fastest-growing minority group is Hispanics, some of whom speak only Spanish. In general, recent immigrants with minimal English will tend to be concentrated in lower-paying jobs, many of which involve handling dangerous chemicals (pesticides, industrial solvents) or potentially hazardous machinery. Warnings need to be useful to them. In addition, products marketed overseas need to have warnings in the language of the country in which they are sold. See Chapter 8 for more information.

Sometimes the problem is that those coming into contact with the product are not intended users at all. For example, what kind of warning would reach children with access to household drain cleaner? Or what sort of label should go on a high-voltage transformer in an area near a playground?

What Must Be Warned Against?

First, let's look at what does *not* need to be warned against. In general, the manufacturer has no duty to warn of open and obvious dangers — that a knife cuts, for example. However, you must be sure that the danger is obvious to the user. A danger that is open and obvious to you, who are thoroughly familiar with your company's products, might be unknown to the user. For example, many people are unaware that burning charcoal emits carbon monoxide gas, which can be deadly in a confined space. Anyone concerned with the manufacture of charcoal briquettes surely knows this, yet every year the news includes reports of people dying while using a charcoal grill in a trailer or closed garage.

Even if the danger is open and obvious, the manufacturer may have a duty to warn if the user might not be aware of the extent or degree of danger. For example, a person using tile adhesive labeled "flammable" probably would not smoke while using the product, but might well not realize the danger posed by pilot lights, especially those in a different room. *Russell* v. *Mississippi Valley Gas* dealt with a person who received severe burns when he was cleaning paintbrushes with gasoline in a storage room attached to a garage. The pilot light of a nearby water heater ignited the fumes.

The manufacturer has a duty to warn potential users of dangers present in the nature of the product in normal use that are not open and obvious. An example of such a hazard is the potential for "kickback" when the tip of a chain saw encounters a hard object. The danger of being cut by a running chain saw is open and obvious; the potential for kickback is not. The manufacturer needs to warn that certain common uses of a chain saw, such as cutting a downed tree into logs, could result in kickback if the tip of the bar accidentally goes into the ground and hits a rock.

The manufacturer also needs to warn against hazards present in "foreseeable uses" of the product. The key word here is "foreseeable." The manufacturer is not required to cover every imaginable misuse of a product. Liability attaches only "when the product is put to uses that it is reasonable to expect a seller or distributor to foresee" (Restatement Third, §2. Comment m). Thus, a manufacturer of chlorine-containing laundry bleach certainly has a duty to warn of potential skin irritation, since this is a risk inherent in normal use of the product to bleach clothes. However, the manufacturer may also have a duty to warn against mixing the product with ammonia (as someone might do if using the bleach as a household cleaner), which produces deadly chlorine gas, because this is a reasonably foreseeable, although not intended, use. On the other hand, a manufacturer of glass casserole dishes may have a duty to warn that a dish taken out of a hot oven and placed on a cold surface may break (reasonably foreseeable), but would not have a duty to warn against breakage caused by using the same casserole dish to drive nails. Again, "reasonableness" is the test.

Post-Sale Duty to Warn

Because the focus in strict liability in tort is on the product, rather than on the conduct of the manufacturer or seller, there is a continuing duty to warn of hazards that are discovered after the time of sale, provided certain conditions are met. The rule is simple: "One engaged in the business of selling or otherwise distributing products is subject to liability for harm to persons or property caused by the seller's failure to provide a warning after the time of sale or distribution of a product if a reasonable person in the seller's position would provide such a warning" (Restatement Third, §10.a). To meet the reasonableness test, four conditions must be met:

1. The seller knows (or should know) of the risk.
2. The users can be identified and don't know the risk.
3. A warning can be effectively conveyed and heeded.
4. The risk is great enough to warrant putting the burden to warn on the seller (Restatement Third, §10.b).

Note that the post-sale duty to warn is a relatively new concept, and the courts have not universally embraced it, particularly when the product was not defective at the time of sale. An example might be a home desk manufactured in the late 1950s with a pullout shelf intended for a portable typewriter. It turns out that, while it is of adequate strength to support a small typewriter, it can collapse under the weight of a computer and monitor. That anyone would have a computer for home use was inconceivable then — prior to the invention of the printed circuit and the silicon chip, a computer would fill an entire room. The desk was not defective at the time of sale, because the use of it to support a personal computer was not foreseeable.

Sometimes the danger is discovered long after manufacture and sale of the product. This can be a particularly thorny problem with prescription drugs or other medical products, in which unexpected side effects can take a long time to appear. In these cases, the manufacturer must keep abreast of current scientific findings. On the positive side, because prescription drugs are, by definition, prescribed by physicians, it is relatively easy for the manufacturer to reach those physicians with a warning.

Much more difficult in terms of finding the users are consumer products, where it is difficult to trace ownership. In these instances, if the risk is serious enough, manufacturers and sellers may publish a warning as a paid advertisement in public media. Naturally, that is not likely to be as effective as a warning conveyed directly to the user, but it may be the only venue available to the seller, short of recall.

A similar issue arises with industrial machinery. This sort of product typically has a very long useful life. A machine may be used by the original purchaser, then resold, and sold again. While the manufacturer may have a continuing duty to warn of newly discovered hazards (providing the users can be found), it does not automatically have a duty to inform users of each safety improvement that has been made since the original purchase. "If every post-sale improvement in a product design were to give rise to a duty to warn users of the risks of continuing to use the existing design, the burden on product sellers would be unacceptably great" (Restatement Third, §10.a).

An interesting sort of post-sale discovery of danger is that resulting from unforeseeable use. You will recall that the manufacturer has no duty to warn of unforeseeable uses — but once that use has been called to the manufacturer's attention, it is no longer unforeseeable. When the plaintiff can show that the manufacturer became aware of a hazard and did nothing to warn users, settlements and verdicts tend to get expensive for manufacturers. For example, in *Krosby* v. *Sukup Mfg., Inc.*, the plaintiff became entangled in a grain auger while he was cleaning a grain-drying bin with the auger running. The grain-drying bin had been purchased in 1973. In the 1980s, the manufacturer began warning about this hazard, but failed to make any effort to notify previous purchasers. The case settled for $2.7 million.

Obviously, the practicality of warning present-day users of products manufactured and sold years ago varies with the product. With a drug, the manufacturer can warn physicians who might prescribe it. With an industrial lathe that has been resold several times, or a household blender, the problem is much more difficult. Clearly, when it is practical to do so, manufacturers would be well advised to keep good customer records. Sometimes the best that a manufacturer can do is to make a good faith effort to warn users of newly discovered hazards or, especially when the danger is great, to offer to repair or replace the hazardous component.

Older products are of particular concern to manufacturers because safe use of older products often depends heavily on the operator's being aware of hazards and taking steps to avoid them. Manufacturers have no guarantee that anyone but the original owner of a product will receive the instruction manual. In addition, safety equipment on older products was often easily removed or bypassed. For example, a recently manufactured drill press might be designed so that the power cannot be turned on unless a movable guard is in place. An older drill press might include the guard, but make its use optional. A manufacturer might be held partially liable for injuries resulting from the (unauthorized) removal of guards because such removal would be considered foreseeable.

DEVELOPING GOOD WARNINGS FOR YOUR PRODUCT

We have said earlier that the best approach is to design hazards out of the product in the first place. If they cannot be designed out, provide a physical barrier (a guard or interlock) to protect against the hazard. You should rely solely on a warning to protect the user only if the hazard cannot be designed out or guarded against. The reason for this hierarchy is simple: the effectiveness of a warning depends on the user's *seeing* it, *understanding* it, and *heeding* it. If any one of those parts in the process goes awry, the user is at risk. The less you rely on warnings to keep users safe, the better.

Except for hazards discovered post-sale, the best time to begin identifying where warnings are needed is when the product is being developed. Once the hazards that must be warned against have been identified, then appropriate warnings can be developed.

Conducting a Hazard Analysis

A good hazard analysis is not the work of one individual — especially not a technical writer. Yet the technical writer needs to be part of the process. A number of different systems are in use for hazard analysis, and we cannot describe them in detail here. However, we can identify the characteristics that make a particular hazard analysis system effective. A good procedure is

- Systematic
- Comprehensive
- Inclusive
- Documented
- Rational

Let's look more closely at each of these.

Systematic

A hazard analysis is used not only to identify hazards, but also to show that the manufacturer made a good faith effort to address the safety issues in its products. A systematic procedure that is performed the same way each time is much more defensible than a haphazard, catch-as-catch-can approach. Ideally, the hazard analysis is an ongoing process that begins when the product is first conceived and continues through the life of the product. Showing that your company has a standard procedure in place for performing systematic product safety reviews will speak well for the company's concern for its customers.

Comprehensive

The hazard analysis should look at all aspects of the product, and should be repeated at each stage of the product's development. After all, when is it easier to make a design change — when the product is still in the design phase or after it is in production? Obviously, the earlier in the process that problems can be found, the better. The analysis should address not only the product itself, but also accompanying literature, including marketing material. Make sure that marketing materials do not conflict with what the manual says about how the product should be used. And make sure the manual is in concert with the product design. Of course, if you find problems with the manual, redesign is the only option — you can't very well put a warning label on the front cover saying, "WARNING! This manual may be hard to understand."

Inclusive

Just as the best designs result when engineers are in close communication with the production, marketing, and service areas of a company, the best hazard analysis includes representatives from all areas. Each person brings the perspective of his or her own training and experience to bear on the product. The hazard analysis team should include, at a minimum, representatives from these areas:

- Engineering
- Manufacturing
- Marketing
- Service
- Documentation (even if Tech Pubs is a subunit of another listed department)
- Legal

In addition, it is a good idea to include at least one person who is new to the product. Just as we suggested a "person-on-the-street" perspective be used in testing the manual, we also suggest that someone *not* familiar with the product be used in the hazard analysis. Those intimately involved with the development and sale of a product may well be blind to issues facing a new user — and yet it is the inexperienced user who may be the most vulnerable to product hazards.

Documented

The work of the hazard analysis team must be clearly and comprehensively documented. You want to create a paper trail that shows your efforts to eliminate as many hazards as possible. While such a record may be subject to discovery in a legal proceeding (i.e., the opposing side may require you to produce these documents), it will show that your company followed a standard, thorough procedure to address product hazards. In other words, you acted in a responsible and *reasonable* way.

Technical writers and engineers with whom we have talked have often expressed concern about whether such a paper trail can be used against them. Our best advice is to develop a standard format for formally documenting the hazard analysis process, and have that stand as the permanent record. With a formal reporting system in place, informal notes from brainstorming sessions and problem-solving meetings need not be kept as part of the file.

Rational

Rarely can a manufacturer address all conceivable hazards associated with a product — and, as we have seen, you don't really need to. What you do need is a rational process for deciding how to deal with the various hazards that present themselves. This requires some sort of risk analysis.

The two variables that need to be balanced against cost considerations are the *likelihood* of injury and the potential *severity* of that injury. An unguarded auger or chain drive located next to where workers push product into the machine is a life-threatening accident "waiting to happen." It is both likely and severe, and obviously needs to be dealt with. On the other hand, an obscure pinch point in the inner workings of a machine, not normally accessible during use and not exerting much pressure if encountered, is both unlikely and minor. It could probably be ignored. Most hazards fall between the two extremes and need to be assigned priorities. A rational system for doing so is essential.

What Is an Adequate Warning?

While it is impossible to *guarantee* that a particular warning will be found adequate in court (juries are unpredictable), certain elements have emerged as necessary for a warning to be considered adequate. At a minimum, a hazard warning must do these four things:

- Identify the gravity of the risk.
- Describe the nature of the risk.
- Tell the user how to avoid the risk.
- Be clearly communicated to the person exposed to the risk.

All four of these elements must be present. A warning against burning charcoal in an enclosed area that reads, "WARNING! Always use with adequate ventilation. Fumes may cause illness," would clearly be inadequate, because it does not tell the severity of the risk (carbon monoxide is deadly) nor tell how to avoid the risk (what's *adequate* ventilation?). Similarly, a warning in English on a bottle of pesticide marketed to growers whose workforce was primarily Spanish speaking would probably be inadequate.

Sometimes the best way to make sure the user gets the warning is to put a label right on the product itself. When you are designing labels for products, you must look not only at designing the warning itself, but also at placement, size, and durability. Place the warning near the hazard, so that it will be seen by the user. The best warning label is of no use if it cannot be seen. For example, a Kentucky court awarded a $266,000 settlement to a farmworker injured when his hand was pulled into a corn picker. A 3 × 5-inch label on the side of the picker warned to disengage the power before cleaning the rollers — but there was no warning label near the rollers.

Be sure the warning is readable from a safe distance. You don't want the user to have to come dangerously close to a hazard just to be able to read the warning! Think about what kinds of lighting conditions are likely when the product is used: will the light be dim (as in a poorly lit basement or factory) or glaringly bright (as in a sunny corn field)? Choose colors and label material accordingly. What is the viewing angle? Will the user be able to look directly at the warning where you plan to place it? If the angle is too severe, the words may not be readable. Try to foresee all the likely conditions of use for your product — and then make your design decisions.

Make sure the label is durable. A carefully designed warning label is useless if it dissolves in the rain, is rubbed off in a few weeks of use, or is totally obliterated by a few smudges of oil or dirt. In choosing the materials you use for your labels, you must consider the material to which the label will be attached, the length of time it will have to last, and the kind of treatment it will receive. A flimsy decal may be acceptable for a computer used in an office, but would certainly not be adequate for a gravel separator used in a quarry. Be sure that your customers can order replacements for damaged warning labels — preferably at no cost.

Standards for Warnings

Many different industries have standards for their products, including, in some cases, for the warnings that accompany products. There are also general standards for warnings that apply to a wide variety of products across different industries.

ANSI

The most commonly used standard for warnings in the United States is ANSI Z535.4-1998, *Product Safety Signs and Labels.* This standard is not required — compliance with it is entirely voluntary. However, it is so widely used that courts may view it as the "state of the art" in warnings, and a manufacturer that chooses to deviate from it should have a very defensible reason for doing so. One reason might be that legal requirements exist for certain warnings to be included with certain products, such as electrical or natural gas appliances. Clearly, if a particular warning is required by law or regulation, you must use it, since failing to comply with legal requirements is *prima facie* evidence of a defective product. You may wish to supplement that warning with an ANSI-compliant one. The ANSI standard calls for each warning label to consist of three parts:

1. A signal word that conveys the severity of the hazard
2. A symbol or pictorial showing the nature and consequences of the hazard
3. A word message further describing the hazard and telling how to avoid it

See Figure 6.1 for sample warning labels that fit the ANSI standard. The standard and its annexes spell out these requirements in detail, but a few comments are in order here.

The three signal words used are DANGER, WARNING, and CAUTION. They must be used consistently to convey a particular level of risk:

- DANGER (white letters on a red background): "Indicates an imminently hazardous situation which, if not avoided, will result in death or serious injury. This signal word is to be limited to the most extreme situations."
- WARNING (black letters on an orange background): "Indicates a potentially hazardous situation which, if not avoided, could result in death or serious injury."
- CAUTION (black letters on a yellow background): "Indicates a potentially hazardous situation which, if not avoided, may result in minor or moderate injury. It may also be used to alert against unsafe practices."

Note that two issues dictate which signal word should be used: the *severity* of the consequence and the *likelihood* of its occurrence. Thus, the decision to use DANGER

⚠DANGER

**Hazardous voltage.
Will cause
severe injury or death.**

Do not open this cover
if blower is running.

See instruction book.

NOTICE

**Connect
thermostat leads
in series
with Stop Button
of 3-wire pilot circuit
in motor controller.**

⚠WARNING

**Hazardous voltage.
Can shock, burn,
or cause death.**

Turn off power
before inserting
maintenance handle.

⚠WARNING

**Hazardous gas.
Can sting eyes,
irritate nose,
or cause death.**

Ventilate breaker
before entering.
See instruction book.

Figure 6.1 Sample warning labels. (From *Product Safety Label Handbook*, Westinghouse Electric Corporation, Pittsburgh, PA, 1981. With permission.)

rather than WARNING has to do with whether the hazard is imminent or only potential, *not* whether the likely consequence is death or mere injury. These signal words are placed at the top of the warning label, and the safety alert symbol (exclamation point within a triangle) is placed next to them.

The pictorial or symbol portion of the label should depict the nature of the hazard (and, if possible, its consequences) in a way that is instantly recognizable. The pictorial should *not* depend on words to be understood — the pictorial is one way to reach illiterate users, including children, and those who do not read the language of the message panel. See ANSI Z535.2 for guidelines on effective pictorials. Figure 6.2 shows a number of typical pictorials used to alert to various hazards. In addition, a number of companies offer catalogs of pictorials and warning labels, as well as custom-design services.* A good pictorial can help solve the problem of the illiterate or non-English-speaking user.

The word message should be simple, direct, and active. Do not be afraid to word a warning strongly and specifically. Do *not* say: "May result in bodily harm"; *do* say: "Can amputate fingers." The most common fault with warnings is that they are too vague. For example, consider this warning:

> This product contains asbestos fiber. Inhalation of asbestos in excessive quantities over long periods of time may be harmful. If dust is created when this product is handled, avoid breathing the dust. Use with adequate ventilation.

Here is the warning again, this time with the questions it fails to answer:

> This product contains asbestos fiber. Inhalation of asbestos in excessive quantities [*How much is that?*] over long periods of time [*How long?*] may [*Possible, not probable?*] be harmful [*In what way?*]. If dust is created when this product is handled, avoid breathing the dust [*How?*]. Use with adequate ventilation [*What's adequate?*].

Obviously, the reader of this message is given very little real information. A better warning might read like this:

> WARNING! Breathing hazard. This product contains asbestos fiber, which can cause serious lung disease if inhaled. Wear an approved respirator when handling. Read and follow instructions before using.

Here, all three parts of the required content are present. The signal word tells the severity of the hazard (potential for death or serious injury). The message explains the nature of the hazard (breathing asbestos dust) and adds more detail on the severity (serious lung disease). The message also tells how to avoid the hazard (wear an approved respirator). The question, "approved by whom?" is addressed by referring the user to the written instructions that came with the product. And it does all that in eight fewer words than the original.

* One of the leaders in the field is Hazard Communication Systems, Inc. P.O. Box 1174, Milford, PA 18337.

Figure 6.2 Sample pictorials. (From *Product Safety Sign and Label System,* 3rd ed., FMC Corporation, Santa Clara, CA, 1980, 7-2. With permission.)

Some manufacturers fear that writing forceful, specific warnings will make their products seem unreasonably dangerous. In general, potential buyers are glad to have specific information, and the courts have held numerous times that a vague warning is an inadequate warning. A successful products liability suit can cost a company a good deal more than the loss of a couple of sales to persons "scared off" by well-written warnings.

ISO 3864

For safety signs and labels accompanying products for export, it is best to follow the ISO 3864 standard.* This standard defines three kinds of signs:

* Products exported to Australia and Canada may use the ANSI Z535 format. Note that products exported to Quebec must have the message in French as well as in English.

- *Warning*: Triangular yellow and black sign with a pictorial denoting a hazard
- *Mandatory Action*: Round blue and white sign with a pictorial to indicate what action to take to avoid a hazard, such as reading a manual
- *Prohibition*: Round black and white sign with the familiar red circle and bar across a pictorial indicating an action that must *not* be taken

Note that unlike ANSI Z535 labels, ISO labels and signs may not contain words at all, and the pictorials used may not always be immediately obvious. However, through repetition and training, users will become familiar with the meanings attached to each symbol. Commonly, the three types of signs are used in combination to convey the full safety message. Sometimes explanatory text is added.

OSHA 1910.144

This OSHA standard governs safety signs used in facilities such as factories and other workplaces. The OSHA standard, when it was first adopted, was based on the ANSI Z53 and Z35. ANSI Z535 replaces both of those, but the OSHA standard has not yet been updated to reflect the change. Nevertheless, developing facility signs that meet the ANSI Z535 standard will certainly comply with OSHA regulation — in fact, the ANSI format is the *preferred* format.

Including Warnings in Manuals

In addition to placing labels on a product, good instructions must also include warnings in appropriate places. Warnings in the manual must be consistent with any warning labels on the product. For example, do not use one signal word on the label and a different one in the manual for the same hazard. The manual will probably contain more warnings than are placed as labels on the product itself, but any DANGER-level warning in the manual should have a corresponding label on the product. Be careful that you do not overuse the DANGER designation — too many of them may dilute the impact. If you find yourself writing an inordinate number of DANGER warnings, it may indicate that the product is unreasonably dangerous and should be redesigned.

Be careful not to embed warnings in the text in such a way that they might be missed. Remember how people use manuals: they skim, looking for the particular information they need at a given time, rather than carefully reading from beginning to end. Buried warnings may not be seen, much less heeded. We recommend formatting warnings in a distinctive way, so that they stand out from the instructions. Using the ANSI Z535 label format (with or without color) is one approach to this. If you use the ANSI Z535 format, be sure that you define the meaning of the signal words at the beginning of the manual. Whether the ANSI Z535 format is used or not, the warning must still contain all four elements of a good warning (nature of the hazard, severity, how to avoid it, and clear communication) to be adequate.

If possible, avoid grouping all the safety warnings together on one page at the beginning or end of the manual. First of all, if they are all on one page, it is all too easy for the reader to skip that page and miss the warnings. Second, if the warnings are grouped together, they will not be in front of readers when they are working through a procedure. We recommend that you put warnings in the text wherever they are relevant — and that you repeat a warning if it applies to more than one situation. Make sure your readers see the warning before they act.

If your company's legal department insists on having a page of warnings at the beginning of the manual, go ahead ... but put them in the text as well, wherever they are needed. If you do have a "safety page" at the front of the manual, be sure to organize it so that warnings about related hazards are grouped and labeled with a heading. And be sure not to overwarn: limit it to hazards that are serious and probable. Too many warnings, especially about trivial or unlikely hazards, dull the senses and dilute the impact of the needed warnings. Do not, for example, warn (as one manufacturer of industrial equipment actually did) against the dangers of drinking hot motor oil.

Safety Cartoons

While safety cartoons are not as common as they once were, we still see them occasionally. They were developed in an attempt to make the warnings contained in them more noticeable, and they are eye-catching. Most people will look at a cartoon in a sea of text. However, in our view, cartoons have a number of serious drawbacks, and we do not recommend their use:

- Cartoons may dilute the warnings in the text — especially if the warnings incorporate pictorials, as we recommend. You cannot include a cartoon for every warning, but including cartoons for only some may imply that the other warnings are trivial.
- Cartoons may seem to treat a heavy subject lightly and thus undermine their own purpose. The cartoon format may remind product users of Saturday morning TV cartoons, in which terrible things happened to the characters without rendering them any permanent harm.
- Cartoons are very hard to design well. To focus attention effectively on a hazard, a cartoon must be extremely simple and uncluttered and must make the hazard itself clear and obvious.
- Cartoons may imply that you are talking down to readers, thus alienating them and making it less likely that they will read (and heed) other warnings or instructions for safe practice.

Given all these shortcomings, and given the availability of effective symbols and pictorials (which are also eye-catching), we recommend that you avoid using safety cartoons.

SPECIAL PROBLEMS

As we have seen, the manufacturer of a product is liable if the product is defective in any way, including in its instructions and warnings. Two aspects of this duty to provide adequate instructions and warnings are especially difficult:

1. How to make sure the user gets the manual
2. How to make sure the instructions and warnings are adequate when the product uses components manufactured by another company

These are difficult areas because they are not entirely in your control. Here are some ideas for dealing with each of these issues.

Making Sure the User Gets the Manual

Getting the manual to the user can be problematic when the user is not the buyer. This is typically the situation when a company purchasing agent buys products that are then used in manufacturing. All too often, the manuals are filed in the purchasing agent's office, and the actual operator of the machine or user of the chemical never sees them. Or the user may have a manual for the product, but it is not the manual that belongs with that model. With machines, one option may be to design a special compartment on the machine itself for the manual. Then ship the product with *two* manuals: one to be filed and one to stay with the machine. Some manufacturers even attach the manual to the machine by a chain or similar device. For other products, such as industrial chemicals, instructions and warnings may be printed right on the container in which the product is shipped. In any case, be sure that the manual is dated and tells exactly what model(s) it covers and what, if any, previous manuals it replaces.

For products sold through distributors, the manufacturer needs a way to ensure that the dealer passes the manual on to the customer. Here some sort of documentation may be required. It may be as simple as requiring the dealer to have the new owner fill out a postcard at time of sale, indicating that he or she received the manual. In some cases, manufacturers include training in the purchase price — and part of the training is going through the manual. When the training is complete, new users sign off that they have received the manual.

The most difficult situation, of course, is when the product is resold by the buyer to someone else. The manufacturer has no way to control this transaction or to guarantee that the manual will be passed on to the purchaser. If possible, the product itself should be labeled with the manufacturer's name, address (including Web address), and telephone number, along with information on ordering the manual.

Products with Components Manufactured by Someone Else

The law extends the duty to have good instructions and warnings all through the distribution chain. Thus, if your company manufactures a printing press in which the controller is made by one company, the rollers by another, and the drive mechanism by a third, you may be liable for inadequate instructions or warnings pertaining to any of those systems. Conversely, if you manufacture the drive mechanism, you have no control over the application in which another manufacturer uses it, yet you may be liable for inadequate instructions and warnings.

Most companies simply include the manuals from component manufacturers (commonly called original equipment manufacturers, or OEMs) with the main product manual. For example, if you buy a power lawn mower, you will often get two manuals, one for the mower and another (from a different manufacturer) for the engine. The difficulty for you, of course, is that you cannot guarantee the quality of the OEM instructions and warnings. Beyond including additional warnings about the component product (which your company may not even have the expertise to write), about all you can do is take comfort in the fact that if you are successfully sued, you can probably turn around and recover your costs by suing the OEM.

SUMMARY

The technical writer is often given the task of designing and writing warning labels for the product and warnings for inclusion in the manual. To do a good job requires familiarity with both legal requirements for warnings and standards that apply in particular industries. This chapter has provided an introduction to this complex topic. We began with a discussion of product liability law and how that affects the manufacturer's duty to warn.

The first step in creating adequate warnings is to conduct a hazard analysis that identifies aspects of the product that may present a risk to users. If it is possible to do so, redesign may be the most effective way to address those hazards. Simply warning against them is always the last resort, but frequently is necessary. A good warning will meet applicable standards, including voluntary ones, like ANSI Z535, as well as international standards. Warnings in the manual must be in concert with warnings on the product itself, and generally should follow the same guidelines, although they present additional challenges.

Finally, we discussed two especially difficult areas: making sure the warnings reach the ultimate user, who may not be the buyer of the product, and dealing with products in which component parts are manufactured by someone else. Both of these areas require creative approaches to ensure that the manufacturer meets the legal duty to provide a safe product with adequate instructions and warnings.

CHECKLIST

- □ Have I conducted a thorough hazard analysis of the product, using a team approach and being sure to look at the product through the eyes of a first-time user?
- □ Have I explored design alternatives and guarding for hazards?
- □ Have I identified and prioritized residual risks according to severity and likelihood?
- □ Have I anticipated foreseeable misuses of the product?

- ☐ Do my warnings include all four required elements?
- ☐ Do my warnings conform to applicable standards?
- ☐ Are my warning labels sized and positioned appropriately?
- ☐ Will my warnings be durable for the expected life of the product?
- ☐ Have I included appropriate warnings in my manual?
- ☐ Are warnings in the manual formatted to stand out?
- ☐ Are warnings in the manual consistent with warnings on the product?
- ☐ Is there a way to document that the buyer received the manual?

REFERENCES

General

ANSI Z535.4-1998. *American National Standard: Product Safety Signs and Labels,* National Electrical Manufacturers Association, Rosslyn, VA, 1998.

Moll, Richard. Oral presentation at the University of Wisconsin Department of Engineering Professional Development, Madison, March 26, 1999.

Peters, G. A. and Peters, B. J. *Warnings, Instructions, and Technical Communications*, Lawyers and Judges Publishing Co., Inc., Tucson, AZ, 1999.

Restatement of the Law Third, Torts: Products Liability. American Law Institute Publishers, St. Paul, MN, 1998.

Case Citations

Hubbard-Hall Chemical Company v. Silverman, 340 F 2d 402 (1st Cir. 1965).

Krosby v. *Sukup Mfg., Inc.,* U.S. District Court, W.D. Mos., No. 89-0110-CV-W-9, Feb. 15, 1991.

Moore v. *New Idea Farm Equipment Co.,* 78-C1-069 (Lee County, Ky. Cir. 1981).

Russell v. *Mississippi Valley Gas,* Miss., Hinds County Circuit Court, No. 31,510, Mar. 20, 1987.

CHAPTER 7

Service and Maintenance Manuals

OVERVIEW

Service and maintenance manuals follow the same design principles as operator manuals, but the resulting documents are decidedly different, because they have a much more specialized audience and purpose. This chapter addresses the design of both service and maintenance manuals. The two forms are similar, but not exactly the same. Generally, a service manual covers repair and maintenance procedures for a single product, often a portable or mobile one. Examples of products for which you would expect to find a service manual include automobiles, pumps, sewing machines, and fax machines. Maintenance manuals typically apply to large, stationary, industrial machines: such things as milling machines, paper converters, printing presses, and turret lathes. These machines often have component parts (such as controllers or motors) that are manufactured by a company different from the one that built the machine. The maintenance manual must include information about maintaining those parts as well.

Certainly, the principles of good verbal and visual design outlined elsewhere in this book will still apply: instructions should be presented in parallel forms, graphics should be clear and easy to read, and so on. However, the application of these principles will differ because of the difference in audience and purpose. The audience for an operator manual can be anyone from a professional user who is technically sophisticated to a member of the general public who has never seen the product before, much less used it. By contrast, the audience for service and maintenance manuals is presumed to be technically sophisticated and, often, very familiar with the product. The audience will most often consist of professional repair and service technicians or industrial maintenance technicians, but it may also include knowledgeable amateurs — the do-it-yourself auto mechanic, for example.

The purpose for an operator manual is to give clear instructions for the use and care of a product. It introduces new users to the product and explains what it is for and how to make it work. An operator manual may give some simple maintenance procedures, such as how to clean the heads on a VCR or how to change the oil in a lawn mower, but such instructions usually cover only the most basic operations.

For anything complicated, the user is usually referred to "an authorized service representative." The service or maintenance manual, by contrast, is what the authorized service representative uses. The purpose of these manuals is to explain in detail the repair and maintenance of a product — to replace the heads on the VCR or overhaul the engine on the lawn mower. Normally, a service or maintenance manual will assume that the reader is familiar with the product, knows how to operate it properly, and needs specialized information only.

In addition, service manuals often serve as "textbooks" for training technicians. In factory training programs and in technical schools, students learn by doing. A student in a transmissions class at a technical college, for example, will learn how to repair auto transmissions by working on one — from a particular make and model of car. The service manual for that car will be the primary resource for the student learning how to do a particular procedure.

What should these books contain? How should they be put together to serve the needs of this specialized audience set. This chapter addresses these questions and provides specific suggestions for making these manuals effective.

EFFECTIVE DESIGN FOR SERVICE AND MAINTENANCE MANUALS

Because the users and functions of service and maintenance manuals are very different from those for operator's manuals, these manuals have a different look and "feel" to them. Specifically, a comparison will show differences in four areas: content and style, format, graphics, and safety.

Content and Style

Unlike an operator manual, a service or maintenance manual will usually not include any introduction to *using* the product. Although it may peripherally address such issues as capabilities of the product or limits of operation, the focus is not on operating it, but rather on installing, maintaining, and repairing the product. Of course, the precise content will vary with the product, but certain categories of information are typically covered.

A *service* manual will generally address these topics:

- Specifications for the product, including capabilities, limits of operation, and capacities
- Recommended lubricants, cleaning agents, and other consumables
- Technical background on the function and operation of the product and its systems
- Procedures for adjustment, operating tolerances
- Routine maintenance procedures and recommended service intervals
- Troubleshooting guide
- Repair procedures
- Model variation information
- Parts catalog (may be a separate publication)

A *maintenance* manual will contain all of the above categories of information, plus these:

- Unpacking/uncrating the machine, including location of lifting eyes or lift points
- Installation, including space requirements, floor strength requirements, and hookup of utility systems (electrical, hydraulic, pneumatic, steam, gas, etc.)
- Special air-handling requirements (venting, exhaust, dust collection)
- Start-up and shutdown procedures
- Schematics of all systems (electrical, hydraulic, pneumatic, etc.)
- Manuals from the manufacturers of component parts (generally called "Original Equipment Manufacturers" or OEM manuals)

There is no set order for these elements, although it stands to reason that Installation should be covered before Start-Up or Troubleshooting. Some of these topics (especially in a service manual) may also be covered in an operator manual, although they will be treated differently.

The information in a service or maintenance manual will naturally be far more complete than that in an operator manual. More complete explanations will be given, and more complex procedures will be described. All of the organizational and writing strategies described in Chapter 3 apply to service manuals as well as to operator manuals. The use of such techniques as general-to-specific organization, lists, parallel structure, and active voice are just as important to the reader of the service manual. The differences in style between an operator manual and a service manual have to do primarily with level of language and tone, rather than with the basic principles of presenting information.

The language in a service manual will be a good deal more technical than the language in an operator manual. Since anyone reading a service manual has a certain amount of technical expertise, the writer can use a more-specialized vocabulary. Don't make things technical just for the sake of making them technical, however; good writing of any sort is as simple as it can be and still convey the necessary information concisely.

Remember also that not all readers of the manual will be familiar with the manufacturer's particular name for things. If you use a term that is special to one manufacturer, try also to include the generic name for the item as well — particularly if you are referring to a tool. It is common practice for factory-produced service manuals to refer to tools by factory numbers instead of generic names (e.g., "VW 558" instead of "flywheel holding fixture"). The writer of a service manual must be aware that *not all* users of the service manual will be authorized, factory-trained technicians. It might appear that using a specialized terminology for specialized tools would discourage amateurs; however, it is more likely that the amateur will simply find some other way to do the procedure (e.g., using a screwdriver rather than a snap-ring pliers). If a specialized tool is *necessary* for a given procedure for safety reasons or to avoid damage to the product, note this fact and refer to the tool by name as well as number.

The pace of a service manual may also be somewhat faster than that of an operator manual. This simply means that you can present information at a faster rate on the page. You may include less background information and more substantive words per sentence. Remember, however, that the emphasis in a service or maintenance manual is on procedures, which means that readers will probably be looking back and forth between the manual and the product as they perform a procedure. Keep your sentences and paragraphs relatively short, and use formatting devices to make it easy for the technician to find the right place in the manual again after looking away for a moment to do a step in a procedure. Overloaded sentences and paragraphs are just as annoying to a technically sophisticated reader as they are to a first-time operator.

Finally, the tone of a service manual may be less conversational than that of an operator manual. As we have noted, the function of an operator manual is in part to represent the company to its customers and in part to introduce the new user to the product gently. This requires that an operator manual be written in everyday language and that it sound "friendly." The purpose of a service manual is primarily to explain to a professional or knowledgeable amateur technician how to perform various repair and maintenance procedures. Example 7.1 shows the differences in tone. The example contains two excerpts, the first from a typewriter owner's manual and the second from a typewriter service manual, both dealing with the operation of the right margin stop. The technology is a little dated, but the principle is as valid as ever.

Service manuals will contain detailed information on specifications and tolerances. Operator manuals will contain some of this same information, but only what is necessary for routine maintenance by the operator. For example, the operator manual for a cassette recorder would include specifications for power input from various sources (batteries, house current, car battery), with specifications for appropriate adapters. A car owner's manual might include specifications for spark plug gap, on the assumption that owners might do their own tune-ups. A service manual, however, would be much more detailed.

Example 7.1 Style differences between operator and service manuals. (Operator manual, from *IBM Correcting Selectric II Operating Instructions*, International Business Machines Corporation, White Plains, NY, 1973, 6. With permission. Service manual, from *Single-Element Typewriter, Model 200, Service Manual*, Silver-Reed America, Torrance, CA, p. 33. With permission.)

Operator manual:

The right margin stop prevents you from typing past the right margin; however, you can space or tab right through it. To type past the right margin, press MAR REL (margin release) and continue typing.

Service manual:

As the Carrier moves to the right still more after the Bell ringing, the Margin Stop Latch moves up the Margin Stop Right extension allowing the Margin Rack to rotate. Then the Margin Plate attached to the Margin Rack rotates the Linelock Bellcrank through the Margin Link. At that time, the Linelock Bellcrank extension moves the Linelock Keylever downward causing its extension to insert into the space between the Keyboard Lock Balls. And the Keyboard has been locked to prevent typing past the Margin Stop Right.

The increased detail typically means that the text contains more numerical information.

Dealing with numerical information such as specifications or tolerances in a primarily textual context presents problems. When you include specification information in the text of technical background and procedure sections, you must be especially careful to avoid:

- Letting the numbers get lost in the paragraphs of text
- Letting the numbers obscure the flow of your description of how an assembly works or how a mechanism should be adjusted

This requires constant attention to how numbers relate to the rest of the text. If you have one or two numbers in a long paragraph of explanatory text, the reader can all too easily skim over them and miss what may be vital information. On the other hand, a paragraph loaded with numbers is terribly hard to read. Generally, more than four or five exact numbers in a paragraph of text is too many. The reader simply cannot keep the numbers straight and often loses the line of thought conveyed in the text.

You may use a number of techniques to solve these problems. To make an occasional numeral stand out in a sea of words, print it in boldface. A manual with different typefaces may be a little more expensive to produce, but is well worth it if the extra expense ensures that the material is used. As an alternative, the numbers may be separated by white space from the surrounding text. If the number occurs in a procedure description, try to put the exact number in a step of its own rather than including it with other adjustments. Example 7.2 shows how this technique can clarify the text.

If the text contains many exact numbers, it may be better to put them in a separate table or chart. A set of exact numbers can be much more easily assimilated in chart form than in paragraph form (see Chapter 5 for more information on tables and charts). The table need not be elaborate or formal. It can consist of simply setting

Example 7.2 Using steps to separate numbers from text. Note how the second version is less confusing.

Original

1. Position the fan motor in the fan housing so that there is 2-3 mm clearance on all sides of the fan and secure with the 4 adjustable clamps. If adjustment is needed, tighten or loosen the clamps one at a time, no more than ¼ turn each time, until the proper clearance is achieved.

Improved

1. Position the fan motor in the fan housing, securing with the four adjustable clamps, maintaining 2–3 mm clearance on all sides of the fan.
2. If adjustment is needed, tighten or loosen one clamp at a time, turning it no more than ¼ turn each time.

Example 7.3 Use of an informal table to separate numbers from text. The second version is much easier to grasp because the numbers are not run in with the text.

Original

When the new extractor is installed, check the fit as follows. Insert a caliper and measure the distance between the hook of the extractor and the opposite side of the breech face. The measurement must be between 7.2 mm and 7.3 mm for the Model 72 and between 6.9 mm and 7.0 mm for the Model 75. If the space is greater than tolerance, file the adjustment pad on the extractor until it is within tolerance.

Revised

When the new extractor is installed, check the fit. Insert a caliper and measure the distance between the hook of the extractor and the opposite side of the breech face. The measurement must be within these ranges:

Model 72	7.2–7.3 mm
Model 75	6.9–7.0 mm

If the space is greater than tolerance, file the adjustment pad on the extractor until it is within tolerance.

the numerical information off from the rest of the text by white space. Example 7.3 shows how this technique improves readability.

The last technique for making numerical information visible is to include tolerance and specifications in visuals accompanying the text. In designing your visuals, you must be careful that numerical information does not clutter up the drawing. A good drawing can easily be ruined with too many labels and excessive specification information, especially in a service manual. Since the audience is generally more technically sophisticated than the audience for an operator manual, the writer may be tempted to use unedited engineering drawings for visuals. Although such drawings contain a wealth of specification information, they are usually too cluttered to be useful for service or maintenance procedures. It is certainly possible to use drawings well to convey specification information as Figures 7.1 and 7.2 show. Note especially the use of the close-up to illustrate a particular portion of the drawing. Be sure that if you use drawings to present specification or tolerance information, that the drawing is placed next to the relevant text, particularly in procedures sections.

Format and Mechanics

As with operator manuals, the needs of the user determine the appropriate choices for format and mechanical elements in service and maintenance manuals. Users of service and maintenance manuals normally need very precise information on how to perform specific procedures. The design of a service or maintenance manual must make it easy to find the precise information needed. Generally, these aspects of manual design may be handled differently in these more technical manuals:

- Organization and referencing
- Page layout
- Binding

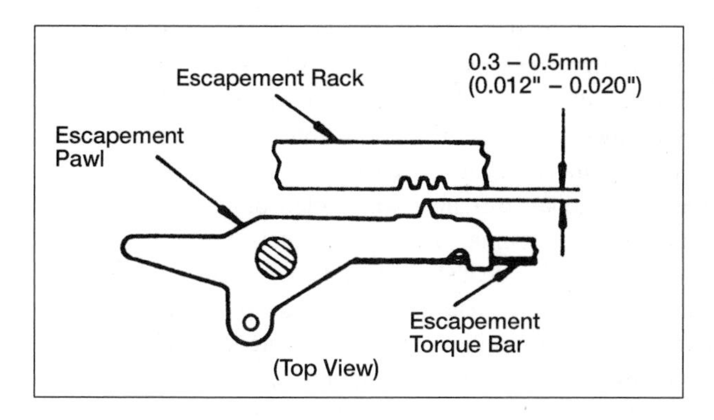

Figure 7.1 Example of a graphic used to convey specifications and tolerances. (From *Single-Element Typewriter, Model 200, Service Manual*, Silver-Reed America, Inc., Torrance, CA, p. 44. With permission.)

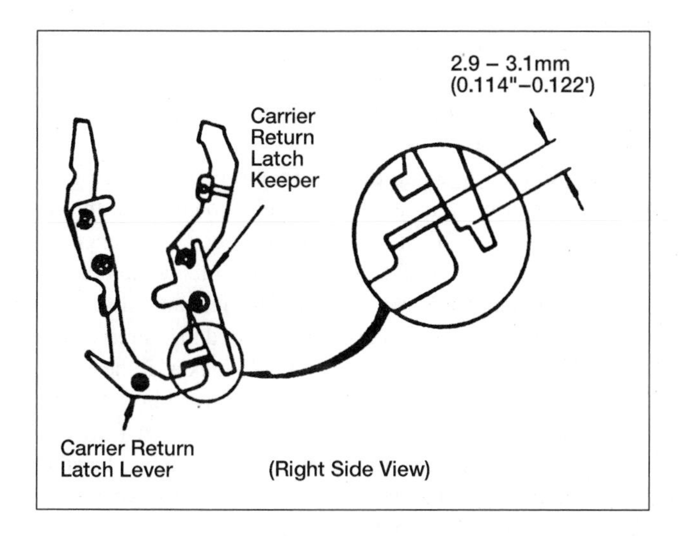

Figure 7.2 Example of a graphic that conveys information about tolerances. Note the effective use of a close-up to avoid cluttering the drawing and to make it more readable. (From *Single-Element Typewriter, Model 200, Service Manual*, Silver-Reed America, Inc., Torrance, CA, p. 45. With permission.)

Organization and Referencing

Organization is included here under format because it is tied with how the user accesses the manual. The organization is still determined by "user questions," but the users of service manuals ask different questions. The new owner of a product is concerned with how the product works, how to care for it, and so on. The service technician or maintenance engineer wants to know how to maintain or repair the product. The new owner will probably flip through the pages of the operator manual, reading more or less at random. The technician or engineer will look for a specific section that covers the necessary procedure and will read that section only, unless specifically directed elsewhere. Thus, a service manual must be organized to help the reader locate the procedure or explanation needed for a particular job and to direct the reader to other relevant sections.

Service and maintenance manuals are often organized by product system: for example, an auto service manual will have chapters on the engine, the transmission, the cooling system, the electrical system, and so on. This kind of organization makes it easy to find the needed information. If a car has a problem with the cooling system, the technician knows to go to that chapter first. Frequently, individual chapters will have their own tables of contents. See Figure 7.3.

This organizational system works very well when the technician or maintenance engineer knows the precise location of a problem. However, the systems of any product interact, and the cause of the malfunction may not be immediately obvious. Even if the user goes to the proper section, other information located elsewhere in the manual may be important as well. Because the manual divided up into sections does not easily show the overlap of systems, the writer must take care to include a comprehensive index and cross-referencing as appropriate. If possible, include cross-referencing within the index, e.g., "carburetor, adjustment (see also fuel filter)." The idea is simply to make the manual useful to readers by making it as easy as possible for them to find the section needed.

The problem of how to organize the information in a service manual becomes particularly difficult when the manufacturer decides to combine the service manual and the owner's manual in one book. Although one book is cheaper to produce than two and ensures that everyone has the same information, we do not recommend this practice. The service manual has a very different audience and purpose than the operator manual, and this difference should be reflected in content, style, and format. To combine both kinds of manuals in one book makes it nearly impossible to maintain the appropriate distinctions. Furthermore, including the service manual with the operator manual may encourage some users to perform procedures they should not perform because they are not skilled enough or do not have the appropriate tools. We believe that it is much better for the manufacturer to keep operator manuals and service/maintenance manuals separate.

Page Layout

Almost all service and maintenance manuals use 8½ × 11-inch pages. These manuals are usually used in a shop or factory, and kept in the technician's or engineer's office. They do not need to be as "portable" as operator manuals, and the larger page size allows for more readable illustrations. Because the 8½ × 11-inch size has become so standard, we do not recommend deviating from it. Odd-sized manuals are hard to file and easy to lose.

The large page size dictates, to some extent, the layout. Certainly, a single-column layout running the full width of the page would be difficult to read. Some technical manuals use a two-column format, but the 2/5 layout is more common. This gives a column width of approximately 5 inches — a very readable length for

SECTION 2: ENGINE SYSTEMS

Table of Contents

Figure 7.3 Sample table of contents for a chapter in a manual. This kind of sectionalizing keeps the main table of contents from becoming too cluttered and is another example of general-to-specific organization. (From *Service Manual, Series 2 Four-Wheel Drive Tractors, Applicability: 1977 Production*, Versatile Farm Equipment Company, Winnipeg, Manitoba, Canada, p. 2-1. With permission.)

10- or 12-point type. In addition, the remainder of the page provides space for illustrations and notes. We recommend including some blank pages at the end of each chapter for additional notes.

As with operator manuals, the typeface should be chosen for easy reading. A font size of 10 to 12 points is ideal. It should not be smaller than 10 points. Even though these manuals are usually used indoors, the lighting may be poor.

Binding

Service manuals may be bound (perfect or spiral bindings are common), but machine maintenance manuals are usually placed in three-ring binders. This loose-leaf format has several advantages:

- New information, such as technical bulletins, can be easily added.
- Corrections or revisions can be sent to users in the form of replacement pages, avoiding the need to reprint the whole manual.
- Related information, such as OEM manuals, can easily be inserted into the same binder.

Whatever sort of binding is used, loose-leaf or otherwise, it must be able to withstand hard use. Where an operator manual may be read once and then filed away to lie untouched unless a problem occurs, a service or maintenance manual will be consulted regularly. Even if a technician or engineer is thoroughly familiar with the product, he or she will need to check specifications and tolerances. Both the cover and the pages will need to be sturdy enough to withstand frequent use in potentially dirty environments. A plastic or plastic-coated cover is usually a must for a service or maintenance manual. We would also recommend somewhat heavier-than-normal stock for the pages of such a manual, particularly if it uses a loose-leaf format.

Graphics

As with the verbal portion of a service or maintenance manual, the basic principles for visual design (explained in Chapter 5) apply to both operator manuals and service/maintenance manuals. Good visual design is perhaps even more important in the more technical manuals because so much of a service or maintenance manual is devoted to procedures for repair and adjustment. The combination of verbal and visual material must make the procedure perfectly clear. This often means that the balance of material shifts toward the visual: a service or maintenance manual will tend to have more drawings and photographs than an operator manual. Although the principles are the same for both kinds of manuals, again their application differs. A service or maintenance manual will have more technical drawings; exploded diagrams and cutaways rather than perspective drawings; and circuit diagrams rather than block diagrams. You must take great care to ensure that these are large enough to see easily and are not cluttered.

An exploded diagram, for example, can often be made much more comprehensible by dividing it into sections. Figure 7.4 shows how "sectionalizing" an exploded

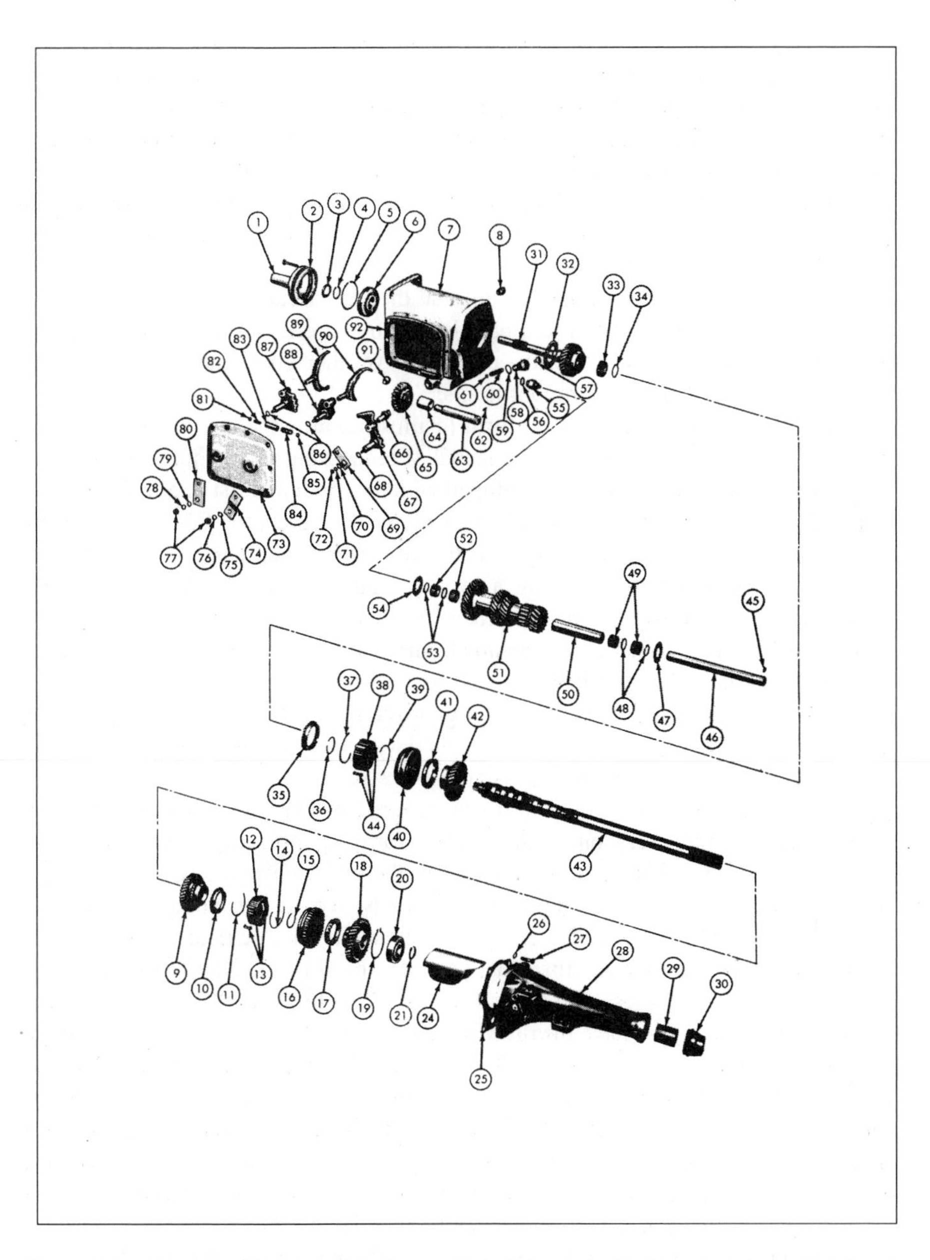

Figure 7.4 Example of "sectionalizing" an exploded view to make it appear less cluttered and to enable the reader to view smaller parts in closeup. (From *Dodge Dart, Coronet and Charger Service Manual*, Chrysler Motors Corporation, Dodge Division, Detroit, MI, 1967, 21-18. With permission.)

drawing of a transmission permits more-complicated portions to appear in closeup. The whole view could have been laid out in one piece and photoreduced to fit on the page, but the result would have been difficult to read.

As suggested in Chapter 5, when the complexity of the drawing permits, label parts with part name rather than a callout. Since the technician or engineer using the manual will already be looking back and forth between the manual and the product, adding another place to look (the key that identifies the callout) will only increase the possibility of a mistake. You must also ensure that the lines showing how parts fit together are easily distinguished from lines or arrows leading from labels or callouts. One good way to do this is to use broken lines for the former and solid, heavier lines for the latter.

If you use cutaway drawings, be sure that the reader can easily differentiate the "layers" of the cutaway. Often you can do this by careful shading — but be sure that your shading does not clutter the drawings. At other times, the best choice may be to use color to highlight different levels.

Perhaps the most easily abused form of illustration is the circuit diagram. The writer of a service or maintenance manual should avoid the temptation to use a reduced version of the schematic developed with the product. First of all, the original schematic was probably drawn on a large scale or designed on a CAD station to be plotted on a large scale. Reducing it to fit onto the manual page would render it unreadable. Second, particularly for a service manual, it may contain more detail than the technician needs, which can lead to unnecessary clutter. Instead, have a schematic drawn or edited for the manual, one that includes only necessary information and is scaled appropriately for the manual page size.

If your product manual includes a circuit diagram, you must be sure that your readers can interpret the symbols used. For a device that is primarily electronic, this is not usually a problem. Someone without knowledge of electrical circuitry is not likely to use a radio service manual, for instance. However, for a product in which an electrical system is only one component — farm equipment, for example — and for which the users of the service manual are likely to be diverse, you may wish to include additional information. For example, the writers of a tractor service manual included the chart shown in Figure 7.5 as explanation for electrical symbols.

For machine maintenance manuals, you may need to include design drawings or complicated schematics. Rather than reduce these to fit an 8½ × 11-inch page, we recommend that you make the drawing the original size, but put it in the manual as a foldout. The easiest way is to place the folded drawing (label visible) in a plastic envelope that is designed to fit in a three-ring binder. Alternatively, the folded drawing can itself be punched to fit into the binding. This has the advantage of making the drawing theoretically more likely to stay with the manual. Practically, however, it is likely to be taken out of the binder for easy viewing, making it all too easy to tear out the holes. A plastic envelope is a better choice.

Good visual design simply means making sure that your visuals are big enough to be easily seen, as simple as they can be while still conveying the necessary information, and clear enough to be easily understood. How these basic principles are put into practice depends on your audience. Since the audience for a service or maintenance manual is likely to be more knowledgeable than the audience for an operator manual, you can use more technically sophisticated visuals — but you should still make them big, simple, and clear.

SYMBOL	MEANING	SYMBOL	MEANING
	Wires crossing. No connection.		Mechanically actuated switch: normally closed, held open.
	Wires connected.		Multi-position rotary switch. Connections and positions as tabulated in diagram.
	Ground connection		Relay, single pole.
	Meter or gauge, as labelled.		Solenoid and Valve
	Motor, DC.		Resistor, fixed.
	Lamp, illuminating.		Resistor, variable.
	Lamp, indicating.		Fuse, current rating as labelled.
	Battery: two or more cells.		Circuit breaker, current rating as labelled.
	Switch, general.		Compressor clutch.
	Pushbutton, normally open.		Speaker.
	Pushbutton, normally closed.		
	Thermostat switch, closes on rising temperature.		
	Pressure sensor, closes on rising pressure.		
	Pressure sensor, opens on rising pressure.		

Figure 7.5 Page of electrical symbols included in a manual to help readers find their way through a circuit diagram. An excellent example of writing with the user in mind. (From *Service Manual, Series 2 Four-Wheel Drive Tractors, Applicability: 1977 Production*, Versatile Farm Equipment Company, Winnipeg, Manitoba, Canada, p. 6-4. With permission.)

Safety

As in the operator manual, safety information is critical in a service or maintenance manual. In some ways it may be even more crucial, because the users of these manuals will be performing more-sophisticated — and potentially more dangerous — procedures. Naturally, all the aspects of including warnings and safety information that are discussed in Chapter 6 will also apply to service and maintenance manuals. In addition, the following subjects may need to be addressed in a service or maintenance manual:

- Lockout/tagout procedures
- Revisions/changes in procedures

- Training of technicians and maintenance personnel
- Compliance with codes

These are rather broad areas, with a good deal of individual variation across industries and products. Nevertheless, a few general comments are in order.

Lockout/Tagout Procedures

For many industrial machines, maintenance procedures require that guards be removed to give access to parts of the mechanism. Obviously, the maintenance engineer or technician adjusting unguarded mechanisms may be at considerable risk if the machine is accidentally started. Some of these machines are very large, and an operator at one end may be totally unaware that someone is working on an unguarded mechanism at the other end. For that reason, a *lockout/tagout* procedure is commonly used. Such a procedure entails physically locking (often with a padlock) switches in the OFF position to prevent an accidental start. A special tag is placed on the machine to alert users that it is locked out.

Many companies have their own lockout/tagout procedures. It is nevertheless a good idea for the writer of a machine maintenance manual to include a recommended procedure in the manual. Information that should be covered includes:

- The location of lockout points on the machine
- What each lockout point controls
- When the machine needs to be locked out
- Who should control the keys to the locks

Some companies put the maintenance supervisor in charge of the keys. This practice puts the burden on that supervisor to be utterly sure that it is safe to turn the machine back on. A safer — although potentially more cumbersome — procedure is to have the individual worker lock out the portion of the machine that he or she is working on, and keep the key until the work is done. That worker then puts the machine back in service. In any case, the manufacturer should provide guidance regarding when and how a lockout/tagout procedure should be used. In addition to this general information, each procedure that requires the machine to be locked out must have a clear warning to that effect at the beginning.

Revisions and Changes

As Chapter 6 discusses, the manufacturer has a continuing duty to warn of newly discovered hazards or changes in recommended procedures, providing that the manufacturer can reasonably be expected to reach the users. Because service manuals and maintenance manuals are, for the most part, used by professionals rather than by the general public, it is easier for a manufacturer to locate them and provide updates. Certainly, factory-authorized service technicians will maintain a relationship with the manufacturer and expect to receive periodic technical bulletins and other updates. Similarly, manufacturers of custom industrial machinery typically have an ongoing relationship with their customers, making it possible to provide them with new information.

Service manuals and maintenance manuals, unlike operator manuals, are normally updated periodically. Product users do not need updated information, since they are unlikely to buy successive versions of the same product. In cases where that does happen — a company, for example, may replace office machines on a regular basis — the manufacturer usually supplies a new operator manual with each product.

This approach is impractical, however, for the service technician who repairs and maintains a manufacturer's product line over a period of time. One technician, for example, may service several generations of a particular manufacturer's small gasoline engines — as well as those of other manufacturers. Especially when year-to-year modifications are relatively minor, it is much cheaper and handier for the manufacturer to supply supplements to an existing service manual than it is to write a whole new manual each time the product changes slightly. The writer of a service manual must keep in mind this need for frequent updates and must design the manual so that updating is easily accomplished.

Similar problems appear when a company's products are predominately custom installations, as may be the case with, for example, industrial packaging equipment or medical equipment. Somehow, the company still has to provide an accurate service manual. Computer technology is making this easier, particularly with a modular manual: one can, with the right system, custom-assemble a manual for each machine. For most companies, this degree of automation is a ways in the future, and more traditional methods are the norm, both for customizing and updating.

Many different techniques are used to update manuals. Some manufacturers supply replacement or supplemental pages to be inserted into the service manual which, of course, requires that the service manual be bound in a loose-leaf binder. Increasingly, manufacturers are using video and computer technology to keep their service technicians informed of changes and new procedures. For example, an auto manufacturer may supply its dealerships with a videotape showing a new service procedure being performed, rather than requiring the dealership to send a mechanic to the factory for training. Or technicians may be provided supplemental information on microfiche. New information may be sent out in the form of floppy disks or CD-ROM. Although all these technological improvements in communication are desirable, the fact remains that not all users of the service manual or maintenance manual have access to the technology to use them. Even if such technology is available, the initial "high-tech" communication should be followed up with "hard copy" — including supplements to the manual. These supplements are usually in the form of individual pages or separate pamphlets.

We recommend replacement or add-on pages for two reasons:

1. If new information is bound right into the original manual, the manufacturer is assured that the technician has the new information. Separate booklets are too easily lost or misfiled.
2. If the new information is bound into the manual at the relevant place, it is more likely to be noticed.

To ensure that the updated pages are used, the writer of a service manual can use the following techniques:

- At the beginning of the manual, tell the reader that supplements will be provided from time to time and explain how to use them. Distinguish between add-on pages, which should follow existing pages in the book, and replacement pages, which require removal of the old pages.
- Provide a page at the beginning or end of the book on which the technician can record the addition of supplemental material.

Number pages clearly according to the pagination of the original manual and build in a means of distinguishing add-on from replacement pages. (For example, add-on pages might be numbered with the page number of the page they should follow, plus a letter — 26a, 26b, etc. — whereas replacement pages would simply be numbered the same as the pages they are intended to replace.)

You may also include in a service manual supplemental information about probable field modifications. This information tells the service technician of ways in which the owner may have modified the product. Although these modifications may not be approved by the manufacturer, they do take place, and the technician needs to know of them. On the other hand, if this information is included in the manual, it certainly demonstrates that the modification was foreseeable. If a particular modification appears often, it may be a signal that a design change is needed in the product. If a field modification produces a hazard, such as the removal of guards or bypassing of interlocks, technicians should be directed to reverse the change and put the owner on notice that the modification is not acceptable. Sales representatives and local dealers are good sources of information for the service manual writer about what modifications may be expected.

Training of Technicians and Maintenance Personnel

At the beginning of this chapter, we noted that service manuals often serve as textbooks for technicians learning to repair products. Maintenance manuals may also be used to train plant personnel in routine cleaning and service procedures for industrial machinery. If you expect that your manual will be used as a training tool, you may wish to specify what sections in the manual must be included in a training program. Some manufacturers provide an outline for a training program and even include test questions. If your manual is used often for training, a relatively formal system of certifying personnel makes sense.

In any case, make certain that the procedures in the manual reflect the best practices. Do not recommend, or even describe, "shortcuts" that could be hazardous. Keep the information in the manual current. Make sure that any photographs and illustrations of persons performing procedures reflect proper practices (protective equipment in place, machine locked out, proper tools used, etc.). The text and the illustrations must not conflict. Remember that images are generally more powerful and memorable than words.

Compliance with Codes

Industrial machines (and some consumer products, such as large appliances) frequently must be connected to utilities and other systems within the plant, such as steam pipes or hydraulic lines. Most manufacturers include in the installation section instructions for connecting to these other systems. As we saw in Chapter 6, failure to meet to a legal or required standard automatically makes a product defective. Codes are required standards. Be sure that your installation instructions meet any national codes. Local codes may be more restrictive, and it is impossible to know what all of them require. Given that variety, it is a good idea to specify in the manual that the installation must comply with all applicable codes, and if the manufacturer's instructions and the code conflict, the code takes precedence.

SUMMARY

We have seen in this chapter that although the same basic principles apply to the writing of both operator manuals and service/maintenance manuals, the application of those principles differs because the two kinds of manuals have different audiences and different purposes. Although both manuals may be written about the same product, they will differ in content, style, and organization. The service manual is written for an audience that is more technically sophisticated and is interested primarily in procedures for service and repair. Therefore, the manual will contain information about the technical background of a product system and about procedures for repair. Since the audience is more technically sophisticated, more technical language and a less conversational style may be used. Because of its specialized purpose, a service manual will be organized to assist the technician in finding the exact repair procedure needed. All the differences notwithstanding, the design procedure for both types of manuals is the same: clearly define the audience and purpose and then arrange and write the material to reflect that definition.

CHECKLIST

- ☐ Have I included complete installation instructions, including instructions for unpacking?
- ☐ Have I included all needed specifications, including tolerances, capacities, and limits of operation?
- ☐ Have I included schematics for all systems?
- ☐ Is there a complete troubleshooting/diagnostic guide?
- ☐ Have I used generic names for tools or parts as well as proprietary ones (or included a glossary)?
- ☐ Have I included cross-references, a comprehensive index, and a good table of contents to make it easy to find information?
- ☐ Have I been careful to set numbers off from text when possible?

- ☐ Does the page design make it easy to use? Is there enough white space? A place to jot down notes?
- ☐ Is there a system in place for providing updates?
- ☐ Are the graphics big enough to see easily? Have I included foldouts for drawings that cannot be reduced legibly?
- ☐ Is there complete safety information, including warnings and lock-out/tagout procedures?
- ☐ Is the binding sturdy and dirt resistant? Will the pages stand up to hard use?
- ☐ Have I included all OEM manuals for component parts?

CHAPTER 8

Manuals for International Markets

OVERVIEW

In the last two decades, international trade has grown in volume and complexity. If you have attended national conferences and trade fairs, you are doubtless aware that "Think Global" has become a familiar slogan and that the economic links between nations are steadily increasing. As a consequence, even small and intermediate-size companies whose markets have historically been confined to the U.S. now find themselves selling computers in Africa, rice planters in Southeast Asia, and trucks in China.

When a company decides to market its products outside the U.S., its manual, along with its service and marketing publications, often may have to be produced in languages other than English. Because English is the predominant language of international trade, competent English-speaking representatives or translators will usually be on hand at the initial negotiation and contract stages. However, when products actually begin to be sold and used in other countries, written materials in the native languages become a necessity.

Those companies already involved in international trade expect to produce their manuals in some or all of the following languages.

Afrikaans	Farsi	Hungarian	Russian
Arabic	Finnish	Indonesian	Serbo-Croatian
Chinese	French	Italian	Spanish
Danish	German	Japanese	Swedish
Dutch	Greek	Norwegian	Turkish
English	Hebrew	Portuguese	

If you are interested in doing business with member states of the European Union (EU) — formerly called the European Community, so you'll see the initials EC in some documents — it is helpful to know that the 11 official languages of the EU are Danish, Dutch, English, Finnish, French, German, Greek, Italian, Portuguese, Spanish, and Swedish.

In this chapter we will discuss some of the special problems with manuals in translation and suggest ways to make the translating job easier and more cost-effective.

NEW STANDARDS FOR INTERNATIONAL MANUALS

Researching applicable standards for your product may take a little time. Know from the beginning that "standards" are not universally applicable and are not law. Compliance at this point is encouraged mainly by economics.

In terms of numbers, particularly economic ones, the incentive to comply with European standards is great: the EU has 15 member nations and represents some 370 million people. "More than 130 countries maintain diplomatic relations with the EU." The EU "is the source of some 50 percent of foreign investment in the U.S.," and "up to 3 million highly paid jobs in the U.S. are due to EU investment." About 20 percent of U.S. exports and 40 percent of U.S. investments go to the EU, making it "one of the top two markets for the U.S." (www.eurunion.org/).

If you choose to enter your products into this global market, remember that your manuals go, too. Some recent directives and resolutions from the EU have explicitly addressed the issue of documentation for products and equipment sold internationally. As you read them, remember that, like the ISO 9000 and 14000 series of quality standards, these EU standards are not law — compliance is voluntary — but, increasingly, are encouraged for companies wishing to sell products across national boundaries.

If you have followed us this far, you already know that entering the world of standards is entering a land of acronyms. Half a dozen groups and dozens of initials make difficulty for someone in search of simple answers. Be patient, start keeping lists of names and initials, and make friends with a reference librarian. In the meantime, we'll try to lighten some of the major areas of acronym overload and help you begin your search.

When you start to look for quality standards that affect various products, and their documentation, you'll find the labeling of international standards reflects two major sources: the International Organization for Standardization with its ISO 9000 and 14000 series, and the European Committee for Standardization/Comité Européen de Normalisation (CEN) with its EN 29000 series.

You will also see two other prefixes, BS and ANSI/ASQ, on standards that correspond to ISO and EN series. BS comes from the British Standards Institution, and ANSI/ASQ refers to the American National Standards Institute/American Society for Quality. ANSI represents the U.S. on ISO committees. You may also come across the QS-9000 standards for quality systems within the auto industry.

Discussion of ISO 9000 could fill a book — actually, it does: many books. We've included a list of addresses/Web sites at the end of this chapter to help you find the appropriate information about quality standards for your product/equipment. Keep in mind that the ISO 9000 series of standards are not law but "are now accepted and recognised in virtually every manufacturing and/or trading nation throughout the world" (Dobb 9).

We focus now on one recent international resolution so you can see the intent and direction of the EU as regards production documentation/manuals.

"Council Resolution of 17 December 1998 on Operating Instructions for Technical Consumer Goods (98/C 411/01)" is of particular interest to the technical writer. This resolution encourages member nations, manufacturers, business associations, and consumer associations "to pursue the objective of making information available to consumers, enabling them to make safe, easy, proper and complete use of technical goods" and "to consider, for example, the possibility of voluntary agreements between manufacturers and consumer associations on the design and content of operating instructions and product labelling and award schemes designed to foster the introduction of state-of-the-art, consumer-friendly operating instructions."

The resolution then lists what constitutes "good operating instructions" in seven areas:

1. Development of instructions for use (including standards, guidelines, usability testing)
2. Content (basic sections of manuals)
3. Separate operation instructions for different models of the same product
4. Safety instructions and cautions
5. Language of manuals
6. Communication of information (clear, precise, user-friendly, particularly for groups such as the elderly)
7. Storage of operating instructions for future reference

So you can see how user friendly the EU documents themselves try to be, and how the standard the EU is trying to encourage for manuals is simply good writing/design advice, we'll take a closer look at two of the above sections: language of manuals, and communication of information. (We reprint the text verbatim but have reformatted the Internet downloaded style.)

5. Language of manuals
   - Consumers have easy access to operating instructions at least in their own official [European] Community language, in such a way that they're legible and easy for the consumer to understand.
   - For the sake of clarity and user-friendliness, language versions are set out separately from one another.
   - Translations are based on the original language only and take into account the distinctive cultural characteristics of the area where the relevant language is used; this requires that translations are done by suitably trained experts who share the language of the consumers that the product is aimed at, and that, ideally, they are tested on consumers for comprehension.

6. Communication of information
   - The communication of information ideally meets the following requirements:
     - It is sufficiently clear and precise.
     - It is correct in spelling and grammar, it uses comprehensible words, it uses active verb forms instead of passive forms wherever possible.

- It avoids unnecessary specialist terms.
- It uses everyday expressions.
- It is consistent in the use of words (i.e., the same term should be used throughout to refer to the same object or action).
- It uses typefaces which avoid any confusion between lower case, upper case and figures.
- It explains abbreviations and accompanies them with clear text.
- It ensures that any illustrations used correspond exactly to what the consumer sees, depict only the necessary information and represent only one new item of information per illustration.
- It ensures that any symbols used correspond to commonly used pictograms, are easily recognisable and always have the same meaning.
- When using a combination of text and illustrations, it chooses one of the two as the main medium throughout.
- It does not confine itself to pictures with no text since that does not ensure clarity as pictures alone may not always be sufficiently self-explanatory.

(From Indications for Good Operating Instructions for Technical Consumer.)

Manuals written for international markets should follow the same principles of good writing and design practiced for any audience. Good design and writing, in fact, often help ease the problems of translation you'll face when you ready your manual for meeting multilingual needs at home or for export.

INTERNATIONAL MANUALS: THE USER SPECTRUM EXPANDS

User analysis is always a challenge because levels of technical competence and literacy, gender differences, and age of the users all have to be considered. International markets compound the challenge. Writers have to think about users in both more and less technologically developed countries, about products designed in temperate North America and sold in the tropics or the desert, and about service and parts replacement in countries where the majority of people may never have owned a car or a telephone.

For each nation where your company sells products/equipment and manuals, consider the following particular aspects of users in that country:

- Literacy levels
- Familiarity with various technologies (computers, electronics)
- Access to service, repair, parts (distance, expense, availability, timeliness, consistency)
- Preference for particular products, or products from particular countries (due to status, quality, other reasons)

PRODUCING TRANSLATED MANUALS

The chief problem of producing good translated manuals is their expense. Translated manuals will be the most expensive per copy of all your publications. Ideally, if your manual needs translation into one or more languages, you should try to create an English original manual that needs a minimum of changes. Companies already involved in multiple-language manual production identify these as key problems.

- Identifying competent translators (human, machine, or combination of human and machine) is difficult. Translators sometimes make promises about accuracy or costs and then fail to deliver a satisfactory translation.
- Translated manuals are short runs, and fewer copies of small-batch runs mean higher costs per copy.
- Labeling on visuals must be redone.
- Nomenclature for certain parts and tools varies from language to language.
- Service and repair systems vary widely in quality and comprehensiveness from country to country.
- Parts identification and replacement become more difficult as the supply line lengthens (the more remote the country and the poorer its infrastructure of roads and railway and air travel, the worse the problem).
- Cultural differences may affect manual use. For example, some countries have cultural prohibitions against performing certain mechanical and maintenance tasks. Rural areas, in particular, may have little familiarity with certain machines and their uses.
- Manuals contract or expand as they are translated into other languages. For example, written text usually swells about 20 to 30 percent in translation from English to Germanic and Romance languages, but contracts in translation from English to Japanese.

Marlana Coe, a human factors specialist quoted in an earlier chapter, reminds writers about this physical dimension to translations — in particular the expansion space. She notes (Coe 19):

> What is only four words and twenty-six characters in English, may well be eight words and fifty-four characters in German. When building in expansion space, the rule is to add 15 percent more space, and then add expansion space as follows:

Number of characters	Additional space
1–10	101–200 percent
11–20	81–100 percent
21–30	61–80 percent
31–50	41–60 percent
51–70	31–40 percent
70+	30 percent

Language

Manuals meant for distribution in Western Europe or in "Westernized" countries are easier for companies in the U.S. to prepare because of shared language (including the language of various technologies). The majority of English words, remember, have their origins in Romance and Germanic languages.

Manuals produced in languages farther from English in structure and conventions are harder to deal with. For example, Russian, Arabic, Greek, Japanese, and Chinese all have their own alphabets or characters. An Arabic manual must be formatted to read from back to front, beginning on the last page, and the lines of text are read from right to left. Most Japanese manual text is arranged horizontally, as in English, but some technical Japanese texts are read vertically.

International manuals also require that measurements and specifications be expressed in metric terms, whose conventions are different from the U.S. system. Other conventions also vary from country to country; for example, the punctuation of numbers:

U.S. 1,000,000.025 vs. European 1.000.000,025

If you are marketing products in Canada, remember that Canada is bilingual. Canadian law requires that consumer publications (i.e., those expected to be used by the general public), such as operator manuals, be printed in both French and English.

If you are marketing products in Mexico or, for that matter, a great many parts of the U.S., manuals should be in Spanish. Bilingual labeling is increasingly desirable in the U.S. for a number of products, especially those used in agriculture (farm machinery, pesticides, herbicides, fertilizers) and food processing, where a large portion of the labor force may speak only Spanish. Your labels and warnings on toxic or dangerous products, and on heavy equipment, always should be in Spanish as well as English.

Remember, too, that the Spanish and French spoken in North America have many differences from those languages spoken elsewhere in the world. This point is important to keep in mind when you look for a translator fluent in Spanish or French — or in any language, for that matter. "Chinese" is more than one language to native speakers, for example. Two people from different parts of China may not, in fact, be able to immediately talk with one another. Keep in mind that all inhabitants of one country may not speak the same language, and one language to an outsider may actually be many to those who know about its subtleties, complexities, and regional differences.

Sometimes companies are able to identify particular ethnic groups as closely associated with a particular trade or industry in a particular location. For example, one U.S. manufacturer of a paint sprayer that, when misused, could be dangerous also recognized from good user analysis that a large percentage of its customers were Greek-speaking contract painters. The company now prints some of its instructions and danger warnings in Greek as well as English.

Nomenclature

Names for parts, processes, and procedures vary from language to language. Sometimes commonly used English words have no real equivalents. In fact, even "British English" and "American English" differ notably. For example, consider the following list.

British	American
Petrol	Gas
Earth (electrical)	Ground
Flex	Wire
Spanner (generic)	Wrench
Bonnet (auto)	Hood
Boot (auto)	Trunk
Dynamo	Generator

And what may be a "shovel" in English may be a "spoon" in another language.

Many "mature" industries, those that for decades have been producing products whose essential features change little from year to year (e.g., plows, automobiles, cameras), have created comprehensive glossaries or dictionaries of routine industry terms. These terms are common parlance; i.e., everyone understands the difference between a knob and a handle and knows what is meant by a socket wrench or tire iron.

These efforts to regularize language and to create standard vocabularies are often referred to as *Simplified English*, and NCR, Kodak, Ford, Caterpillar, and IBM are some of the companies that have developed these systems. Other companies, such as John Deere, have internal style guides that specify the technical terms for parts that writers may use in documents. Writers are not allowed to create their own versions.

In looking for consistency, you may be surprised to discover how inconsistent and idiomatic your manuals really are. One editor discovered that manual writers had used seven different terms for a small stop valve on a diesel engine, calling it spacer, washer, shim, stop, intermediate plate, plate, and stop valve. Another writer had used the idiomatic "mule drive" instead of "90° angle belt drive" and another had advised, for fiberglass repair, "Take a piece of rosin the size of a walnut." (In many countries, the walnut example would be meaningless.)

In translated manuals, you should make an effort to standardize the vocabulary for parts and processes and to be consistent in describing routine procedures. For example, do not instruct the user to "oil" the machine at one time and to "lubricate" it the next, or to "monitor the needle for pressure variations" the first time and to "check the gauge" the second. (Note that the terms *oil* and *lubricate*, as well as the terms *monitor* and *check*, actually have slightly different meanings in English, but are sometimes used loosely or interchangeably.)

Standardized vocabulary is also important in processes and procedures. The following are some commonly used procedural verbs:

tighten	remove	raise	press
loosen	add	lower	release
fill	check	attach	stop
empty	place	fasten	start
clean	turn	adjust	replace

Make decisions about the procedural words you will need to use, and stick to them. Record your choices in your writers' and editors' style guide. Again, don't, for example, say "lower the arm" in one place and "allow the arm to drop" in another, or "turn the wheel to the right" and then "rotate the wheel clockwise."

Culture

When new or unfamiliar products are introduced to a country, cultural patterns may significantly affect their use. Ask international representatives about their experiences in the field and they will quickly regale you with anecdotes. You will hear about eggs fried on electric irons (although this happens in college dorms in the U.S., too), refrigerators used as air conditioners (just leave the door open), wood fires built in the cavities of gas ovens, and tire rims mounted backward.

One frustrated farm equipment representative found that the only way he could convince rural workers in a remote part of a country whose economy could seldom afford technologically complicated equipment that a chopper was a dangerous machine was to toss a hapless farmyard cat into the mechanism. (An excellent speaker from the same farm equipment company, with whom we have taught at seminars, gets branded an insensitive hater of animals every time he tells this true story. He's not. The sad truth of the demonstration is that it worked when nothing else had. One would like to think a first-rate manual would have saved the poor cat.)

Such anecdotes reveal difficulties to be found in cultures unfamiliar with technologies that you, writing in the U.S. for a technologically sophisticated company, might take for granted. In addition, the social structure in some countries may be much more rigidly hierarchical than in yours, with clearer divisions of labor among classes or castes of people. Thus, one worker may be allowed to drive a vehicle, but not to change a tire, and fixing a machine may be considered demeaning or socially taboo.

In brief, if your product is being marketed in a country whose culture is markedly different from yours, then your manuals must be crystal clear and as simple and graphic as possible. Don't accidentally become the insular and insulting "ugly American." Do not confuse unfamiliarity with inability: don't underestimate the speed and skill with which other countries learn new information and new technologies.

Coe lists ten specific areas that can cause you problems when you write for international audiences: passive voice, contractions, noun strings, metaphors and cultural references, graphics and color, grammatical and syntactical structures, vocabulary and usage, special characters, humor, and expansion space (Coe 17–19). Before you write your manual for another country, then, do a little research on culture and customs: talk with sales and service people for that area, visit with people who have lived there, check out library books on business customs and travel in that country.

Graphics

The most important component of manuals in translation is graphics. When graphics are clear and self-explanatory, they also help to diminish whatever errors or mistranslation might creep into the translated verbal text.

All of the suggestions for good photographs, drawings, and charts offered in earlier chapters hold true for translated manuals. Graphics should be big, simple, and clear. They should be carefully coordinated with verbal text and planned so that the most important elements are visible (Figure 8.1).

Graphics are particularly crucial when user literacy levels are likely to be low. If you find that, on an average, you are devoting less than half of your manual to graphics, try to add more illustrations. Do this even if, to you, the mechanism or procedure seems perfectly obvious. We recommend, too, that you pay special attention to safety labels and warnings, which should have clear pictorials as part of their essential message and design.

You can grasp the importance of graphics by imagining your own reaction to a manual originally written in a language you don't understand and poorly translated into English. Your lifeline to assembly, use, and maintenance of the product will be the photographs and drawings.

We suggested in Chapter 5 that you use labels directly on the graphic whenever possible. Manuals for translation are a notable exception, largely because of the costs. Key numbers and accompanying legend in English can easily be translated by changing the legend only. For graphics in translated manuals:

- Follow the suggestions for graphic effectiveness in Chapter 5.
- Add more graphics if verbal text exceeds graphics in space allotment.
- Use key numbers with a legend to simplify translation of complex graphics.
- If possible, incorporate graphics or pictorials into safety warnings and labels.

Choosing Translators

Natural language is rich, slippery, and laden with nuance. In the passage from one language to another, the meanings of words are sometimes skewed or miss the mark entirely. For example, an English manual which read, "Secure the ⅝-in. bolt," was translated into German as, "Put the ⅝-in. bolt behind bars." The word "secure" was completely misunderstood. In another manual, "hydraulic ram" became "water goat."

Sometimes mistranslation merely produces howlers, like the computer translation of "The spirit is willing, but the flesh is weak," which the computer translated as "The drinks were abundant, but the meat was rotten." However, mistranslation becomes serious business especially when safety and precision are at issue. Dangerous procedures or processes requiring great precision need to be explained accurately, with no slippage in meaning.

Figure 8.1 Manual with stand-alone visual pages and "half-size" format. These pages rely entirely on the visuals and on symbols used widely in international manuals (OK, X, ✓, and the safety alert). The horizontal format and "half-size" of the manual are designed for glove compartment storage in a truck cab. (From *TS 60686*, Mack Trucks, Inc., Allentown, PA, 1986, 14, 18, 19. With permission.)

Human Translators

Throughout the U.S., many firms and individuals offer translation services. Finding translators for European markets is relatively easy, but competent translation into non-European languages may take some hard searching. Avenues for finding translators are language and engineering departments in colleges and universities, the Yellow Pages in telephone books of metropolitan areas, large companies whose products are similar to yours and who have been selling "international" for a long time, the Internet, and the good old-fashioned grapevine, or word of mouth.

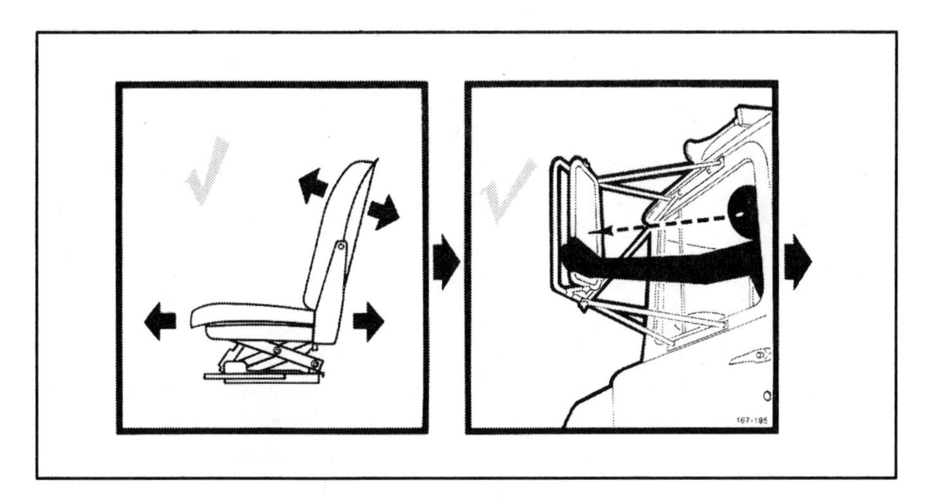

Figure 8.1 (continued).

Many companies have found that their own dealers and international representatives can serve as translators or can identify competent translators within the country. The use of company dealers and service representatives as translators is especially valuable because they know the products as well as the language. As one European put it, "I'm not much good at a social event, but I do know tractor English."

Within the U.S., a great deal of manual translation is a kind of cottage industry. Small firms may be, in reality, single individuals who have created a network of consultants specializing in various languages; many of them work at home. The quality of translation varies a great deal and is, of course, dependent upon individual translator skills as well as upon the complexity of the product for which you need the manual. As you look for individuals or firms to help you, be sure to ask if they have experience with manuals.

There is no fail-safe method for assuring competent and accurate translation, but those with experience in the field offer the following advice.

- Use native speakers in the home country of the language, if their translating skills are coupled with good knowledge of the product.
- Within the U.S., the translator should not have been away from the home language for more than 5 years. Over time, translators begin to lose touch with current idiom and expressions in the native language. They begin to substitute "Englishisms" and American idiom.
- If possible, the translator should have a good working knowledge of the technology of the product or, at the least, a natural feel for how things work. A translator whose specialty is 17th century poetry may be a whiz with the language, but a poor choice for a manual for a circuit breaker or a road paver.

Be prepared to pay for language skills. Some translators charge by the word and others charge by the page or job, depending on the nature of the original. Spending $50,000 or more on a translation is not unheard of. Most of the costs for translated manuals, of course, are in that first copy. If you must rely on a translator who knows nothing about your product, make sure that a dealer or service representative who knows the language checks it for accuracy.

Machine Translation

The computer age has provided translators with a vast array of tools, such as databases for foreign language alphabets and vocabularies, dictionaries and indexes for technical terms, metric conversion, and graphics and formatting options, to name but a few. Many large corporations have developed extensive, industry-specific machine translation (MT) systems. A number of them have ongoing foreign language research projects under way, attempting to speed up and regularize the translating job.

Companies that use outside MT firms report mixed results. Some of the translation systems have been found to have vocabularies too limited for the product. This is often the problem for programs you can now buy for home use. These programs are often limited to one way, English to Spanish, German, or French. Others have found that syntactical and grammar problems in particular languages could not be overcome. Manual producers with long experience using MT say that, as of now, the promise of 80 to 100% accuracy in computer translation has, in many cases, been unrealistic, and that total accuracy in MT is still in the future.

At least two Internet sites, babelfish.altavista.digital.com and Globalink's Comprende at www.lhs.com, offer translations in seconds of small blocks of text. Although these sites are not suitable for company manual translation, they do let you get a feel for the problems and the possibilities of MT. (See Budiansky for details on Internet MT sites.)

Companies that use MT of manuals say it does work very well in the following situations:

- Simple text, limited vocabulary, heavily visual material
- Tabular information, graphs, charts
- Modular format

So far, MT continues to contend with the the unruliness of natural language. If computer translation produces an 80% accurate manual for you, you will still have to rely on human translators for that last 20%. It is a 20% job that many professional translators avoid.

One translator commented, "Being given a computer translation is like being presented with a house that looks like a house, until you get inside. Then you discover that the plumbing leaks and the wiring short circuits. I'd rather move out, start over, and build the house myself."

Choosing the best MT for your product involves as much trial and error as does the choice of human translators. You will need to balance the trade-offs among cost, accuracy, and time, using MT just as you use the other tools of your trade.

Service, Repair, and Parts Replacement

People in the U.S. are accustomed to convenient and readily available service and parts replacement. In international trade to industrialized countries, the advent of modern inventory control and computer update has produced vast improvement in the delivery system of service and repair.

Graphics are again of crucial importance in translated service manuals and parts lists. Many advanced systems of manual production have developed parts catalogs and repair manuals that consist almost entirely of graphics and numbers. Parts lists contain only an identifying picture or drawing and an identification number, and service manuals are done almost entirely with photographs of the product, photographs of the tools needed to assemble and maintain the product, and a limited "universal" vocabulary.

Figure 8.2 shows two pages of a translated manual. One page is the English version and the other is the Finnish version. Notice these features:

- Vocabulary is standardized (choke, valve, stop, start).
- Chart layout makes reading easier.
- Symbol, symbol name, and symbol meaning are clearly laid out and reproduced identically from one language to the other.
- English version is on page 3; Finnish version is also on page 3 of the Finnish section.

Packaging the Translated Manual

Modular format is an invaluable tool in translated manuals. Manuals are frequently packaged in parallel columns, with English on one side and the translation on the other, or with a shared visual format, as shown in Figure 8.3.

When a manual is relatively short and simple, you can make one manual serve for three or four languages by experimenting with foldouts and shared graphics. For example, you can devote the left page to a photograph or drawing of the product and the right page to a foldout with French, German, and/or Spanish text laid out identically and keyed identically to the shared visual. Figure 8.4 shows how a short manual for translation can be formatted to ease translation and keep costs down.

The English version of the manual is 18 pages long. The international manual has the following features:

- One international manual, bound as a single book, serves for ten languages (English, German, Italian, Spanish, Dutch, French, Norwegian, Swedish, Danish, Finnish).
- Each language has a separate section (i.e., a French section, a German section, etc.).
- Page layout, the number of pages, and the page numbering system are identical for each section.
- Shared graphics can be opened out to use with any of the language sections because legends and callout numbers are identical.

Symbol	Symbol Name	Meaning or Purpose of Symbol
	"Functional Description" Symbols	
	CHOKE	Identifies CHOKE control.
	VALVE	Identifies a control valve.
STOP	STOP	Identifies STOP SWITCH control. May also identify STOP position of throttle control on certain motors.
	START	Identifies position of throttle control device during starting. May also identify STARTING control.
	START-MOTOR	Operating device for starting motor.
	"Instructional" Symbols	
	LATCH	Identifies device provided to LATCH or UNLATCH engine cover.
	FUEL SHUT OFF	Identifies device provided to cut off fuel supply to engine.
	SPARK ADVANCE	Number (in degrees) following this symbol indicates recommended maximum spark advance for engine. (Symbol and number appear on engine surface.)
	KEROSENE	Indicates KEROSENE is to be used or identifies KEROSENE is present.
	FUEL	Indicates GASOLINE is to be used or identifies GASOLINE is present.
	OIL	Indicates OIL is to be used or identifies OIL is present.
50/1	FUEL OIL MIX	Identifies FUEL/OIL mixture for 2-stroke engine. Indicates each 50 parts of gasoline are to be mixed with 1 part of oil. Mixture to be mixed completely.

Figure 8.2 Translated manual and standardized vocabulary and symbol explanation (English and Finnish). The chart layout and the one-word system for symbol names plus symbol explanation make translation easier. (From *Evinrude® Outboards*, Outboard Marine Corporation, Waukegan, IL, 1983. With permission.)

VERTAUSKUVA	VERTAUSKUVAN NIMI	VERTAUSKUVAN TARKOITUS TAI VAIKUTUS
	"TOIMINTASELOSTUS" VERTAUSKUVAT	
	KAASUTIN	Osoittaa kaasuttimen saato
	VENTTIILI	Osoittaa saatoventtiili
STOP	PYSAHDYS	Osoittaa pysahdyskytkin Voi myoskin osoittaa kaasuvivun pysahdysasento
	KAYNNISTYS	Osoittaa kaasuvivun asento kaynnistaessa Voi myoskin osoittaa kaynnistyssaato
	MOOTTORIN KAYNNISTUS	Moottorin kaynnistyslaite
	"OHJEITA ANTAVIA" VERTAUSKUVIA	
	SALPA	Osoittaa laitteen joka lukitsee tai irroittaa moottorin kannen
	POLTTOAINEKATKAISU	Osoittaa laitteen joka sulkee moottorin polttoainesyoton
	SYTYTYKSEN SAADIN	Numerot (asteissa) jotka seuraavat tata vertauskuvaa suosittelevat maksimaalinen sytytyksen saadin moottorille Vertauskuva ja numerot ovat moottorin pinnassa
	PALOOJY	Nayttaa etta palooljya on kaytettava tai identifioi etta polttooljy on olemassa
	POLTTOAINE	Nayttaa etta bensiinia on kaytettava tai identifioi etta bensiini on olemassa
	OLJY	Nayttaa etta oljya on kaytettava tai identifioi etta oljya on olemassa
50/1	POLTTOAINE-SEKOITUS	Osoittaa polttoaine oljy sekoitus 2-tahtimoottorille Nayttaa etta joka 50 bensiinin osa pitaa sekoittaa 1 osalla oljya Sekoitus on oltava taysin sekoitettu

Figure 8.2 (continued).

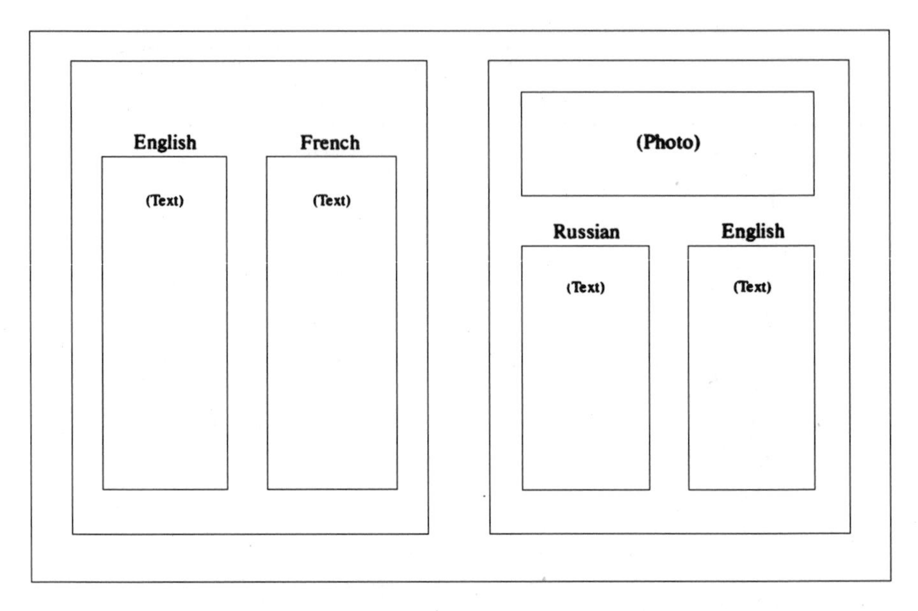

Figure 8.3 Sample layout of multilingual manual, showing "shared" format. The two sample pages in this figure show typical multilingual formats. The prose text is often laid out in parallel columns (e.g., English and French) which allows use of single photographs or graphics that are shared by the various columns of text.

Figures 8.5 through 8.9 show a number of examples of creative packaging for international manuals. Please note that the original pages of Figure 8.6 are in vivid colors. Each language has a designated and unvarying color.

SUMMARY

Companies are increasingly called upon to produce manuals that will be translated into two or more languages. Whenever manuals must be translated, remember these guidelines:

- Standardize your vocabulary.
- Show rather than tell about the product and procedures.
- Use modular format.
- Package creatively.

Writers are often pleasantly surprised to find that writing for translation also improves the original-language version of the manual. Planning ahead lets companies create documents that are easily understood in almost any language.

Incorrect
OVERLOAD FORWARD
CAUSES BOAT
TO "PLOW"

Incorrect
OVERLOAD AFT
CAUSES BOAT
TO "SQUAT"

Correct
BALANCED LOAD
GIVES MAXIMUM
PERFORMANCE

Safety Warning: If engine is tilted forward so as to cause plowing (see A), swamping may occur in rough water. If engine is tilted aft so as to cause porpoising (see B), steering may be erratic or unstable. See correct angle adjustment (see C).

Lubrication

TYPES OF LUBRICANT		Contact your DEALER for OMC Lubricants.	
OMC TRIPLE-GUARD™ GREASE		OMC HI-VIS® GEARCASE LUBE	OIL (S.A.E. 30)
TUBE	GREASE GUN		
A	B	C	D
LUBRICATION PICTURE SYMBOLS			

LUBRICATION POINTS [30] [31] [32] [33] [34] [35]

30. GEARCASE LUBRICATION

Remove oil drain/fill and oil level plugs from side of gearcase. With motor in normal running position, allow oil to drain completely.

To refill, place tube of OMC HI-VIS® Gearcase Lube or equivalent in drain/fill hole. If OMC HI-VIS Gearcase Lube is not available, OMC Premium Blend Gearcase Lubeor equivalent can be used as an alternate. With motor in normal running position, fill until lubricant appears at oil level hole. See **Specifications** for gearcase capacity.

Install oil level plug before removing lubricant tube from oil drain/fill hole. Drain/fill plug can then be securely installed without oil loss.

If the proper tube or filler type can is not available, install drain/fill plug. Slowly fill gearcase through oil level hole allowing trapped air to escape. Install plug.

A. Oil Level Plug
B. Oil Drain/Fill Plug

Change after first 20 hours of operation and check after 50 hours of operation.
Add lubricant if necessary.
Drain and refill every 100 hours of operation or once each season whichever occurs first.

Note Note: Recommended lubricants which have been formulated to protect against damage to bearings and gears must be used as extensive damage can result from improper lubrication.

31. Idle Speed Adjusting Knob Shaft, Spark Advance Linkage, Cam Roller, Shaft and Gears
32. Swivel Bracket, Engine Cover Latch Shaft
33. Shift Lever Shaft and Detent, Choke and Carburetor Linkage
34. Clamp Screws, Tilt/Run Lever Shaft, Tilt Shaft, Steering Handle, Throttle Shaft and Gears
35. Steering Handle Throttle Gear and Bushing

Frequency of Lubrication	
TYPE OF USE	FREQUENCY
Fresh water	Every 60 days
Salt water	Every 30 days
Storage of 30 days or longer	Before placing in storage

Figure 8.4 Translated manual with shared, foldout graphic and standardized format and vocabulary. Note the following features: layouts for Italian and English versions are identical (as are the other six language sections included in this manual); page numbering systems are identical (makes cross-referencing from one language to another easier); the numbers in boxes (30, 31, 32, 33, 34, 35) are lubrication points and correspond to numbers 30 to 35 on the foldout, shared graphic. (From *Evinrude® Outboards*, Outboard Marine Corporation, Waukegan, IL, 1983. With permission.)

⚠ **Avvertimento di Pericolo: Se il motore è inclinato troppo in avanti, la prua si affossa nell'onda e si rischia di imbarcare acqua. Se il motore è inclinato troppo indietro. La barca picchia e la guida diviene incerta od instabile. Cfr. corretta regolazione dell'angolo.**

Lubrificazione

TIPO DI LUBRIFICANTE		Rivolgersi alla CONCESSIONARIA per i lubrificanti OMC	
OMC™ TRIPLE-GUARD GREASE		"HI-VIS"® GEARCASE LUBE OMC	OIL (S.A.E. 30)
TUBO	SIRINGA DI GRASSAGGIO		
A	B	C	D
SIMBOLI GRAFICI PER LA LUBRIFICAZIONE			

PUNTI DA LUBRIFICARE 30 31 32 33 34 35

30. PER SCARICARE LA SCATOLA INGRANAGGI

Togliere i tappi filettati di scarico/rifornimento e di livello a lato della scatola ingranaggi. Con il motore in normale posizione di corsa, lasciar defluire tutto l'olio.

Per rifornire, imboccare il tubo di "HI-VIS® Gearcase Lube" OMC od equivalente nel foro di scarico/rifornimento. Se il lubrificante "HI-VIS Gearcase Lube" OMC non fosse reperibile, si potrà ripiegare sul "Premium Blend Gearcase Lube" OMC o suo equivalente. Con il motore sempre nella normale posizione di corsa, riempire finchè il lubrificante sale a lambire il foro di livello. Per la capacità della scatola ingranaggi, cfr. Specifiche.

Montare il tappo di livello dell'olio prima di staccare il tubo del lubrificante dal foro di scarico/rifornimento. Si potrà così riavvitare il tappo di scarico/rifornimento senza perdita d'olio.

Se non si dispone del tubo o della siringa adatti, montare il tappo di scarico/rifornimento. Riempire lentamente attraverso il foro di livello, permettendo all'aria di fuoruscire. Riavvitare il tappo.

A. Tappo livello olio
B. Tappo per scarico/riempimento olio

Cambiate l'olio dopo le prime 20 ore di funzionamento, quindi verificare il livello ogni 50 ore. Se necessario, rabboccate.

Cambiate l'olio dopo ogni 100 ore di funzionamento o, comunque ad ogni stagione.

Note Nota: Bisogna usare i lubrificanti consigliati che sono stati formulati per la protezione dei cuscinetti e degli ingranaggi dato che l'uso di un lubrificante non adatto può arrecare danni notevoli.

31. Per il minimo regolare il pomello dell'asse, l'anticipo di accensione la camma cilindrica, l'asse e gli ingranaggi
32. Cavalletto di brandeggio ed alberino della leva di fissaggio della carenatura del motore
33. Alberino e dente di arresto della leva comando cambio; articolazione dello starter (comando gas) e del carburatore
34. Viti di chiusura, ingranare e disingranare la leva dell'asse, disingnare l'asse, la leva di direzione, l'asse e i comandi della farfalla
35. Ingranaggio comando gas e cuscinetto della leva di direzione

Frequenza di Lubrificazione	
TIPO DI UTILIZZAZIONE	FREQUENZA
Acqua dolce	Ogni 60 giorni
Acqua di mare	Ogni 30 giorni
Rimessaggio di almeno 30 giorni	Prima del rimessaggio

Figure 8.4 (continued).

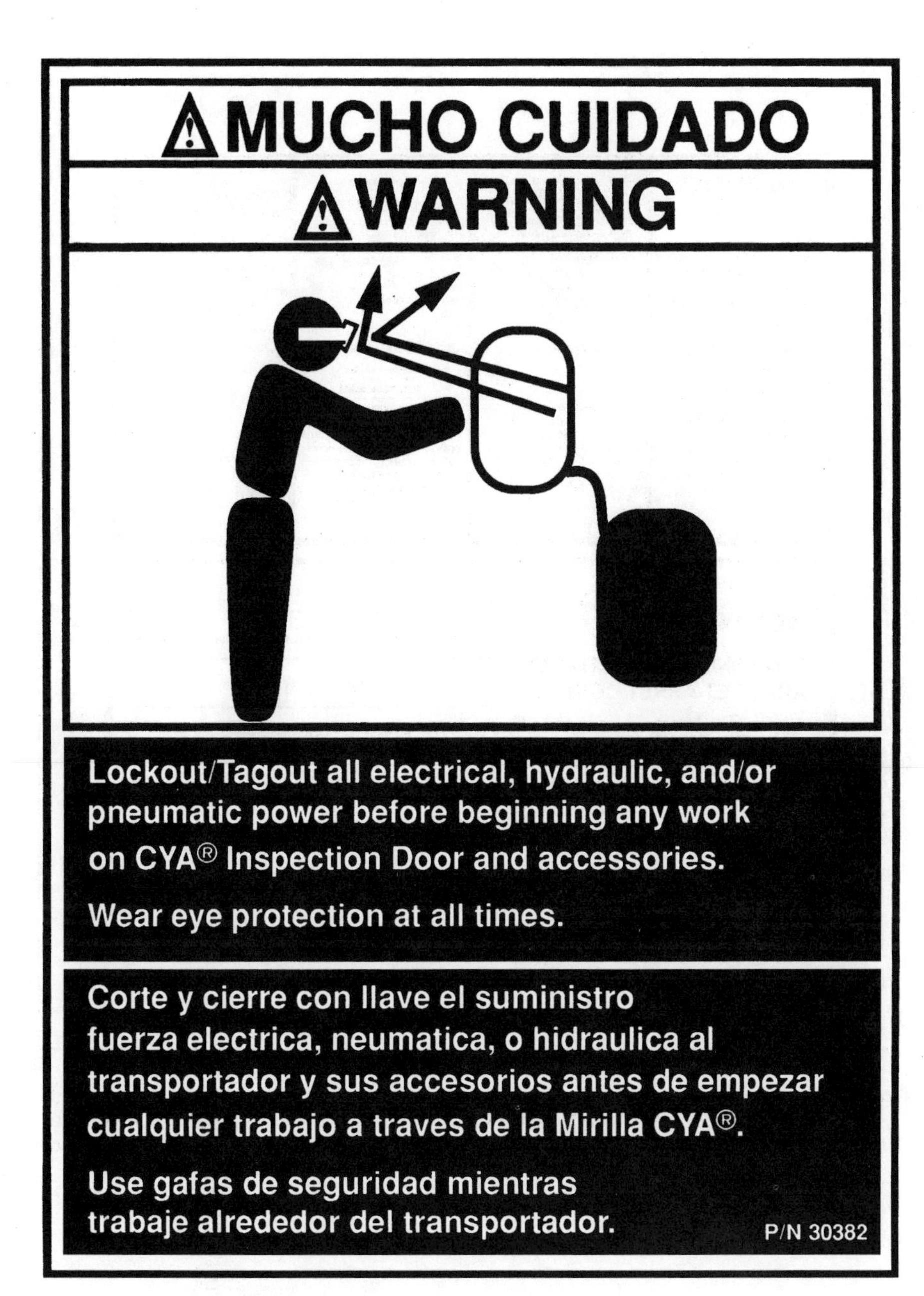

Figure 8.5 Bilingual warning label. The label uses orange for level hazard (Warning), a pictorial, and bilingual presentation of all prose text. (Courtesy of Holly A. Webster, Martin Engineering Company, Neponset, IL, 1990.)

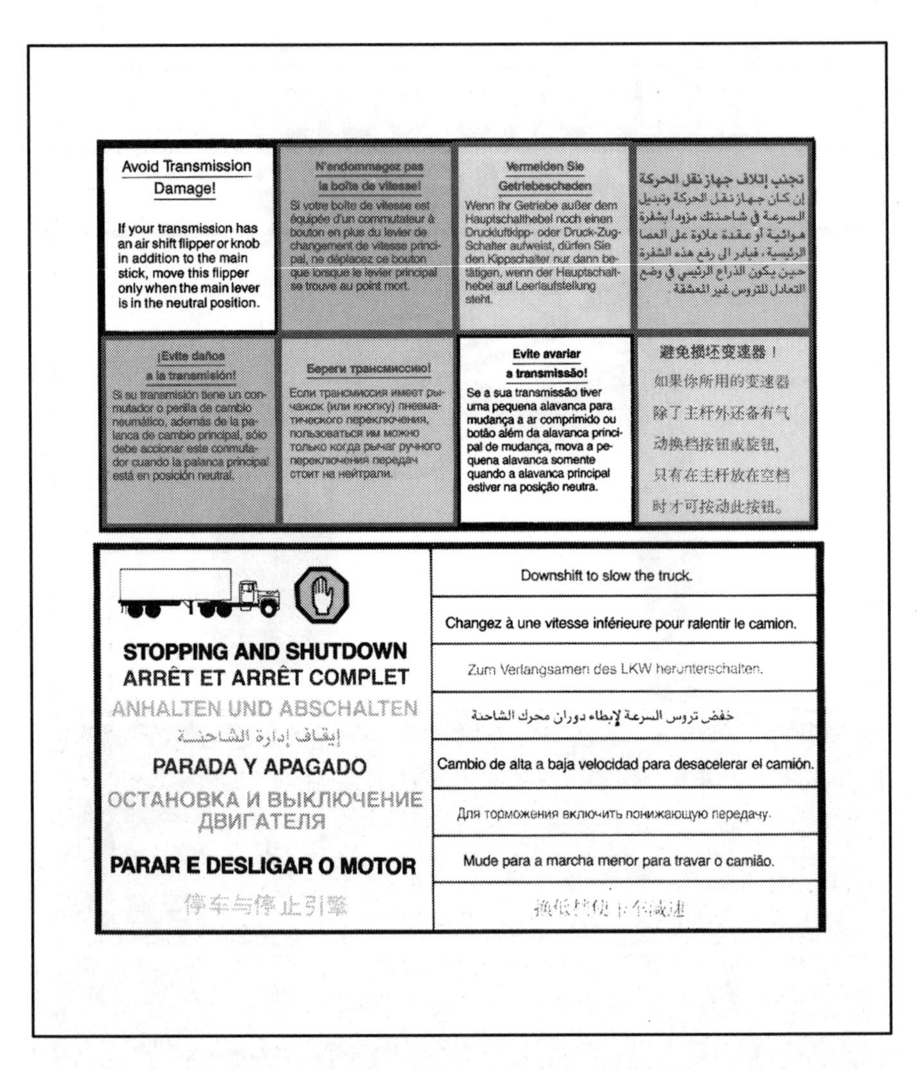

Figure 8.6 Translated manual using color and modular format. Colors are used consistently throughout the manual which serves for eight languages. Modular format requires that text be kept short and simple. (From *Operator's Manual, TS 60691*, Mack Trucks, Inc., Allentown, PA, 1991, pp. 32–33. With permission.)

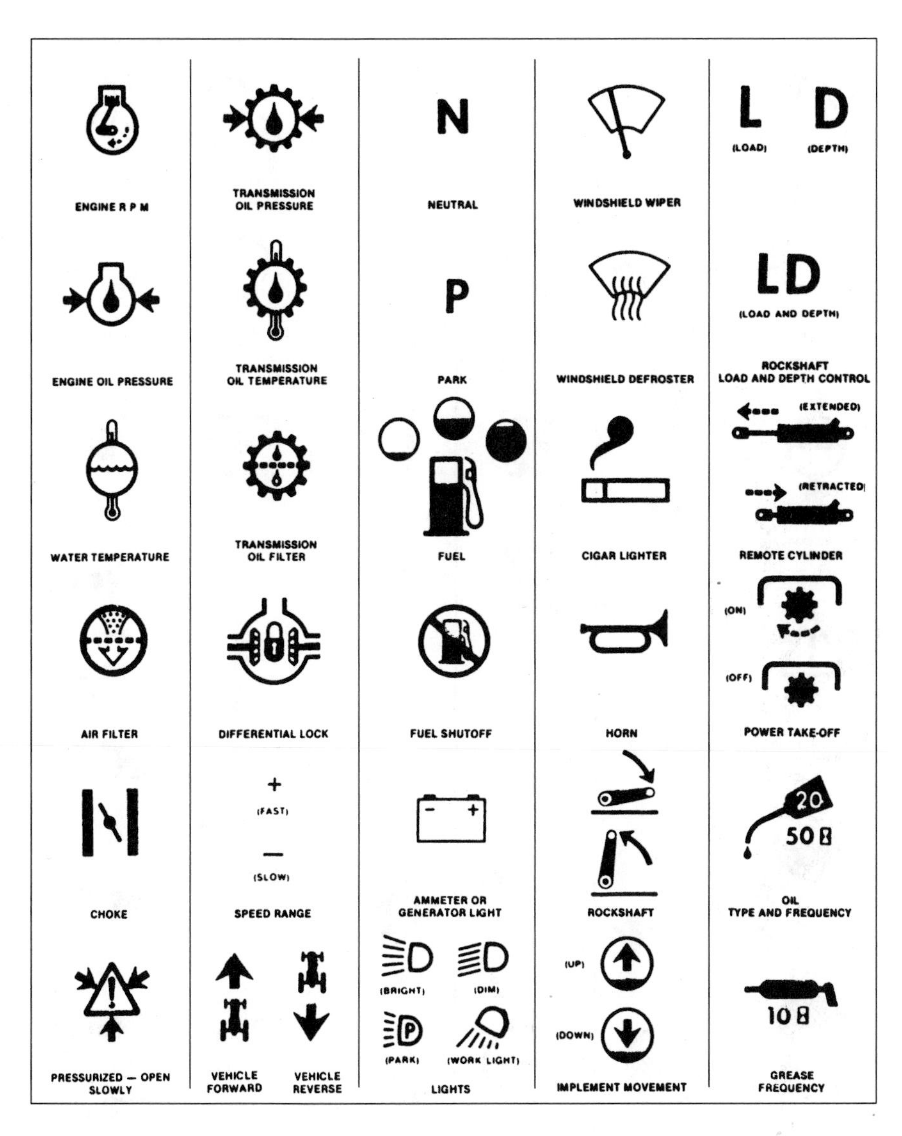

Figure 8.7 International symbols for operator controls. These symbols are standard for automotive and agricultural equipment sold internationally. (From *TD8-44/S4 Grain Augur, PU5003/1800*, Ford New Holland, Inc. With permission.)

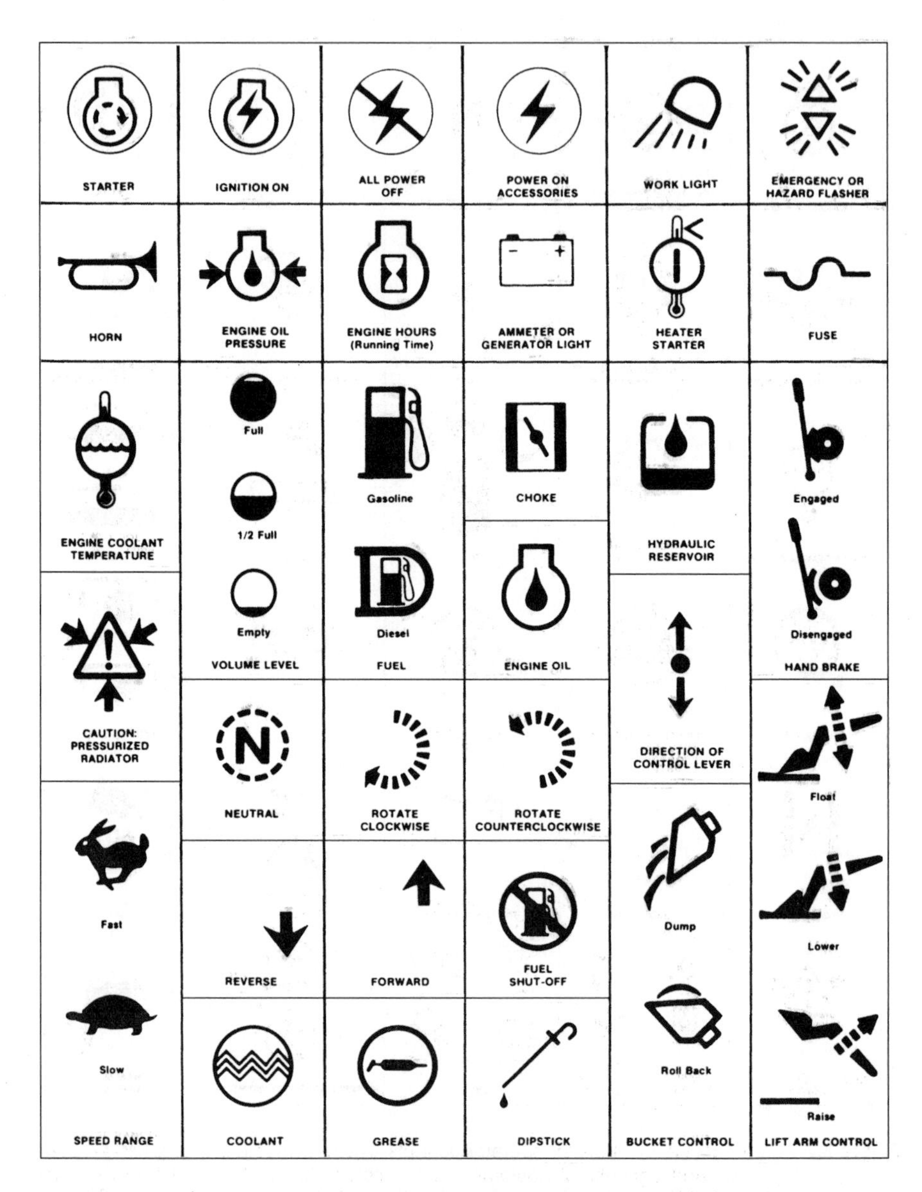

Figure 8.8 Explanation of standard international symbols. These symbols can be used on a wide variety of mechanical products. (From *Skid Loader 4510/4610 Operator's Manual*, Gehl Co., West Bend, WI, p. 28. With permission.)

Controls and Indicators

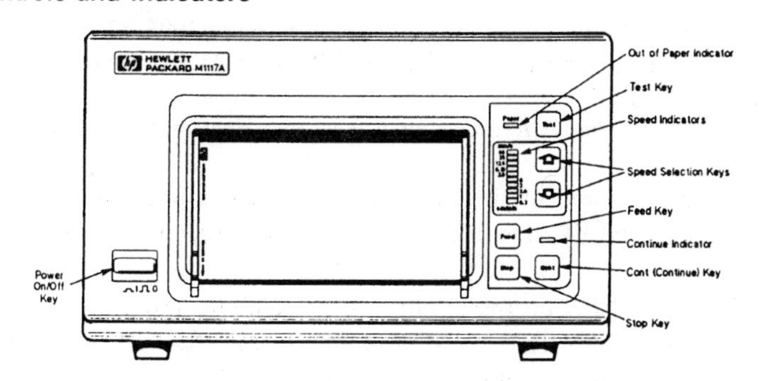

Figure 1-6. M1117A Controls and Indicators

Power On/Off Key	Press this key to turn the recorder power on and off.
Out of Paper Indicator	When the LED is lit, it indicates that the recorder is out of paper or that the compartment door is open.
(Test) Key	Press this key to reset the processors and initiate an internal test consisting of ROM and RAM checksums and link integrity.
Speed Indicators	Ten LED's indicate the current speed setting of the recorder.
Speed Selection Keys	Use the up arrow to increase the paper speed and the down arrow to decrease the paper speed.
(Feed) Key	Press this key to advance the paper at 25 mm/sec. Paper will advance as long as key is depressed. If the recorder was running when (Feed) was pressed, it will return to its original speed after the key is released.
Continue Indicator	The LED indicates when the current recording has been made continuous.
(Cont)inue Key	Press this key to make the current recording continuous.
(Stop) Key	Press this key to stop the current recording.

Note An international version of the front panel overlay is also available and will be shipped to certain countries.

International Versions

The following items vary according to the country in which the order originated. Appropriate items are supplied for each country.

- Front Panel Overlays (English or International)
- Power Cords
- Fuses

Note Line voltage switches are set at the factory according to the requirements of the country in which the order originated.

Figure 8.9 Special-order translations. Manuals in English are sometimes coupled with special-order translations for product accessories such as templates for keyboards, and so on. (From *Model M1117A Multichannel thermal Array Recorder*, Hewlett-Packard, Waltham, MA, 1989, 1-7–1-8. With permission.)

REFERENCES

Budiansky, Stephen. Lost in Translation, *Atlantic,* December, 1998, 81–84.

Coe, Marlana. Writing for Other Cultures: Ten Problem Areas, *Intercom,* vol. 44, no. 1, January, 1997, 17–19.

Dobb, F. P. *ISO 9000 Quality Registration,* Butterworth-Heinemann, Oxford, 1996, 1998.

The following are some of the Internet sources for information about ISO, CE marking, the EU, and EU standards. From any one site you can link to many more.

- From the CELEX database: http://europa.eu.int/celex/
  Or from EUDOR, the European Union Document Repository: http://www.eudor. eu.int/

  CELEX has over 200,000 documents in each of the 11 languages of the EU. Access through CELEX or EUDOR will give you the full texts of EU documents. (Be aware that CELEX is often slow.) However, you can only access CELEX by paid subscription or, for free, through a library that is a EU depository. Contact the academic/university or law library nearest you and ask if it is a depository, or look up the list of U.S. depository libraries at the web site: http://www.eurunion. org/infores/index.htm).

- http://www.newapproach.org/directiveList.asp
  This site gives the New Approach Directives for various products.

- http://www.cenorm.be/faq.htm
  This site gives EU information.

- http://www.eurunion.org/
  This site gives EU information.

- http://ts.nist.gov/ts/htdocs/210/216/ir5122.htm
  This site gives the document on ISO 9000 by Maureen Breitenberg, Standards Code & Information Program, TESIS (Technology Services Information System), from National Institute of Standards & Technology, Office of Standards Services.

- http://www.asq.org/
  American Society for Quality, 611 E. Wisconsin Ave., P.O. Box 3005, Milwaukee, WI 53201-3005, 1-800-248-1946, fax 414-272-1734, e-mail: Contact ASQ.

CHAPTER 9

Managing and Supervising Manual Production

OVERVIEW

Manual production cannot be treated in isolation from the company structures that surround it. This chapter is written to help those who are setting up a manual production operation for the first time and those who realize that their present setups do not seem to be working as well as they should.

The quality of service publications depends on a number of key factors: the initial choice and training of writers, the structure and managerial philosophy of the company, the clear delineation of lines of authority, and the fulfillment of writers' basic needs. We include this chapter because we know that the techniques and suggestions we have made to manual writers cannot be effectively applied unless the fundamental company structures are well designed. Good management exists to make it possible for people to do their jobs well.

WHO WRITES THE MANUAL?

Before we begin a seminar on manual writing, we analyze our participants by asking them to fill out a personal information sheet listing their experience and training for manual writing. Here are just a few of the answers:

Engineering graduate (all kinds)
Service and parts manual writer
English or journalism graduate
Law, business, or psychology graduate
Technician
Prototype builder
Company owner
Son of company owner

Transfer from marketing or advertising
Magazine science writer
No experience

Clearly, the entry into specialized technical writing is sometimes through the "front door," but far more often it is an outgrowth of other job duties, a discovery or tapping of a special talent in midcareer, a tangential assignment, or a deliberate second career choice. In small and intermediate-size companies, manual writing is often an add-on to many other job tasks, and writers may be given little guidance on how to proceed.

Choosing the Technical Writer

When a company decides to assign the manual-writing task or to hire a new writer, it often asks, "Should we choose technicians and engineers and then teach them how to write, or should we choose professional writers and teach them the technology of the product?" Posing the question this way can be misleading because of the underlying assumptions — that technicians and engineers can't (or won't) write and that trained writers will probably be technically naive or ignorant.

A better way to think about the choice is to choose someone who *can* communicate and *likes* to. Certainly, some lawyers and engineers are so bound up in jargon that they find it almost impossible to simplify a message for a general public audience, and some wordsmiths write clean and explicit prose, yet are so technically inept that they cannot grasp the workings of the simplest machine. Nevertheless, there are individuals who possess a combination of the necessary communication skills, and these make the best technical writers. They have technical sense about how things work and either know the product from experience elsewhere in the company or can, with a minimum of explanation and hands-on practice, quickly grasp a new technology or product. They recognize clear, correct prose and can also write it. They have a visual sense about drawings, photographs, and format devices, and, as we pointed out in Chapter 1, they have "people" skills.

As a manager, you can devise appropriate screening devices to identify good communicators. For example, you can give applicants a simple device or product, along with basic information and relevant photographs or drawings. Then ask them to write and lay out a sample page of a manual. Another method is to give applicants sample pages of manuals. Choose a spectrum (some good, some bad, some average) and ask applicants to rank the pages and to justify, in writing, the reasons for their choices.

Training for Writers

In Chapter 1, we discussed the writer's two basic needs: information and time. In this chapter, we add a third: training. Most of the technical writers we talk to are eager to do competent work. They are also quick to sense whether management seems to be "for" or "against" them. As the supervisor of manual production, you should be your writers' chief advocate in insisting on information access, time, and training.

Writers need training for manual production, just as they do for any other kind of specialized publication. We have found that some companies provide no training whatsoever or give their writers only the sketchiest of orientations, whereas others have systematic and comprehensive training programs.

What Kind of Training is Best? — The best training programs are those provided before manual writing begins, as part of the planning process. After orientation in the fundamentals of manual production, writers are then given periodic training in special skills. In the long run, providing preliminary training is more cost-effective than waiting until writers are floundering. It may then be too late to correct errors, after considerable money has already been spent on such items as art, photography, or printing.

Companies that have been producing products for a long time, especially if they are also large companies, often have sophisticated and comprehensive training programs for manual writers. If you work for such a company, you may already have had orientation sessions, hands-on practice working with other, more experienced writers, and close contact with your publication managers and editors. If your company has no such training program, here are some of the training techniques you might consider:

- Have new writers work through a manual from start to finish with an experienced writer.
- Give new writers in-house handbooks and style guidelines or workbooks to orient them to company procedures.
- Have managing editors work closely with new personnel in the first months on the job.
- When companies are decentralized and manuals are produced at several places, assign one manager to coordinate quality control of the manuals. (Some companies use a "roving editor" who travels among the various manual production locations.)
- Bring writers together for periodic training sessions on such special work topics as metric conversion, the writing of safety warnings, photographic techniques, or desktop publishing.
- Make libraries and files of company and competitors' manuals available for writers to look at.
- Teach writers "incrementally" by assigning only small segments of a manual for their first assignment, so that they gradually expand their skills and techniques.

Consultants and Seminars — If your company is too small or has too few writers to make in-house training sessions cost-effective, you can consider using periodic outside consulting help. Many smaller companies make use of the continuing education conferences and seminars conducted by universities and technical institutes.

Working with consultants and attending 2- or 3-day seminars gives writers a chance to learn and exchange ideas with other professionals, to bring themselves up-to-date on product liability, and to practice their writing skills. These comments, for example, came from writers who had been given outside help, either through consulting or at a conference:

> This is my father's company, and I got the manual-writing job. After attending this conference, I'm going to redo the whole thing. It scares me to see how many mistakes I've made. If someone got injured or killed with our equipment, we wouldn't stand a chance in court with our safety warnings. We're small. We could be wiped out with one lawsuit. (From a writer for a small, independent, industrial crane firm.)
>
> I can't believe that I've been doing this job for over 3 years, and I never knew the range of choices I had for visuals. I didn't know that a manual could be used as legal evidence either. Why didn't somebody tell me? (From a writer for a large medical equipment company.)
>
> I decided, after this seminar, that we ought to try videotape. I wrote the bible (the script), we hired outside for filming, and my boss was pleasantly surprised. Two tapes that are pretty good for a first try.

Other Helps for Writers — If your company makes no provision for organized training sessions, you can still ease the writing process a great deal by creating style handbooks, writer guidelines, and fact sheets listing steps involved in the manual process. Managers or editors can provide such books for their writers, and solo writers can create their own handbooks to systematize the procedures they plan to use. For example, one company fact sheet, given to writers before work begins on the manual, contains the following information:

- Product name and number
- Deadline dates for completed manuscript in rough form
- Format specifications (column and page width, margins, type size, specifications for photographs and drawings)
- Schedule and locations for viewing the product and for hands-on practice with mechanisms of the product
- Style guidelines (average sentence length, vocabulary and language level, use of active-voice verbs, etc.)
- Notifications of what the other segments of the manual will be and which writer is assigned to those segments (especially important for cross-reference work or for machine systems that interact)
- List of phone numbers and names of people who can provide information and list of key meetings for product development
- References to materials on file that might be reused

In addition, do not neglect the influence of the physical workplace. Noticeable improvements in writers' effectiveness may come with attention to such mundane details as adequate lighting, up-to-date workstations, and ergonomic chairs. We know of one situation in which the time to produce a manual was cut significantly simply by relocating the art department to the same floor as technical publications.

SCHEDULING AND MONITORING DOCUMENT PREPARATION

If writing manuals for constantly changing products is difficult, scheduling and tracking these writing projects is even worse. Managers of technical publication departments find themselves often faced with multiple projects, shifting deadlines, tight budgets, and limited staff. In such circumstances it is hardly surprising to find many managers operating in constant crisis mode.

Manual-writing projects are difficult to schedule for the same reasons that manuals are difficult to write: you don't have control over time and information. In addition, the manager has to work within a budget that is dictated from above. Further compounding the difficulty is that some managers operate under the misconception (sometimes fostered by writers) that writing is a fundamentally different sort of activity from engineering or manufacturing and therefore cannot be planned, scheduled, and monitored using the same methods. Such managers may assign a writer a project and a deadline, but then have no way of assessing progress — unless the writer comes and says he or she cannot meet the deadline. By then, of course, it is too late to add staff to the project conveniently and the department is back in crisis mode.

How to Schedule a Documentation Project

In our view, writing manuals is essentially similar to any design activity, with predictable inputs and outputs that can be set up as a series of milestones or laid out in a Gantt chart. Scheduling and monitoring a documentation project not only permits a manager to make adjustments before a project is way behind, but it also permits building a track record that can help in estimating future projects.

Planning a documentation project requires that you answer three basic questions:

1. What do you need to produce?
2. How much time do you have to work with?
3. What personnel can you use on the project?

These seem obvious, but it is surprising how many managers do not take the time to look systematically for the answers. Let's look at each question.

What Do You Need to Produce?

Before you can begin to do any realistic scheduling, you need to have a very clear idea of the nature of the documentation project itself. Are you writing a user's guide or a service manual? Or both? Are you working solely in paper documentation or will you also be responsible for producing or coordinating a videotape? Identify as clearly as possible all the types of documents or the multiple purposes of a single document.

When you know what you are going to produce, begin to plan the documents in more detail. Of course, most of the time this planning activity will be a team effort between manager and writers, but whoever does it, it still needs to be done. This detailed planning stage involves the following kinds of activities:

- Estimating page count for the final document (including front matter and back matter)
- Preparing a detailed outline of the document
- Estimating graphics requirements
- Identifying tasks required to complete the project (writing, editing, interviewing subject-matter experts, preparing graphics, designing page layout, etc.)

When this stage is finished, you should have a pretty clear idea of the size of the project. It is time to try to fit the project to the time available.

How Much Time Do You Have?

Typically, you will be working under a deadline imposed by someone else, such as the shipping date of the product — determined by marketing. Your job will be to work backward from that deadline to the present to find out how much time is available. Here is the procedure:

1. Backtrack from that shipping date however long it will take for typesetting, printing, binding, and packaging the manual. This is your real deadline.
2. Get out a calendar and count the working days (no weekends or holidays) available between now and then and total them for each month.
3. Multiply the number of days by 6 to find the hours available if you put just one person on the job (use 6 instead of 8 to permit time for meetings, responding to phone calls, and so forth — and adjust the multiplier as experience dictates).
4. Compare the hours available with your projected page count. From experience you probably have some idea of how many hours it takes to produce a page of final copy (including first draft, second draft, editing, and graphics). Typically, companies see a range of values, depending on the complexity of the material and the experience level of staff. Anywhere from 4 to 10 hours per page is pretty common.
5. Use the projected page count and hours available to determine how many staff you will need to assign to the project. For example, if your page count is 150 pages and it takes 10 hours to produce a page, you will need to have 1500 hours to do the book. If you only have 500 hours available, you will need to assign three staff members to work on the project full time in order to meet the deadline.

What Personnel Can You Assign?

Unfortunately, you will seldom have a full set of writers, editors, graphic artists, and assistants available for assignment. They will all be in various stages of working on other projects. What you have to do, of course, is pull somebody off a project that is nearly finished to get the new project rolling and then add others as they become available.

This kind of juggling act is the "tech pubs" manager's principal activity. It may not be quite as stressful as being an air traffic controller, but it shares the element of needing to keep track of a dozen different things at once. The more systematic you can be about assigning tasks and estimating project requirements, the more smoothly the work will flow and the more likely you will be to meet the deadline. To be systematic, you have to have good information about who is doing what and how far along they are — and that is where project monitoring is essential.

Monitoring an Ongoing Project

Monitoring a project is simply keeping track of how far along the work is and comparing that to the plan, rather like measuring actual expenditures of funds against a budget, and then adjusting for variances. However, how do you know how far along the work is? What are the appropriate milestones? One company uses the following rules of thumb:

- The first draft will take 60% of total time.
- The second draft will take 20% of total time.
- The remaining 20% will go to editing, project management, coordination with graphics, and other support activities.

You know already how many hours you expect the project to take. If you have used up 30% of the time and have only one fourth of the first draft done, you probably need to make some adjustments.

Project Logs

Keeping track of the project requires that the writers and others working on a project log their time and activity. Most companies have some sort of weekly time sheet on which employees show how many hours they spent on which project. Normally these are used to allocate costs among different projects. A time sheet used in conjunction with a brief narrative report of specific activities can give the manager all the information he or she needs to tell whether a project is moving on schedule or not.

Of course, the same information (total project hours, hours per page, etc.) can be used for cost estimating: just multiply hours by the salary (plus overhead) for each employee assigned to the project.

But Isn't This All a Lot of Record Keeping?

The kind of scheduling and tracking that we discuss here does take some time to set up and keep current. There are computer tools to make it easier, ranging from spreadsheet programs like Lotus 1-2-3 or Excel to software designed for project management. Whatever aids you use, it still does take time. Is it worth it? The answer to that depends on your situation. If you are a "one-man band," you probably do not need a complex system to track your work. If you are the manager of a 15-person publications department, you probably do need some system to stay on top of progress and recognize the need for adjustments before a crisis develops. If your situation is somewhere in between, you will have to balance the benefits of having the information against the costs of the time it takes to run the system.

The sort of scheduling and monitoring system that we have described does have one additional advantage regardless of the size of your operation: it allows you to build a set of data that you can use to make your estimates more accurate for future projects. If you keep track of the hours per page that a user's manual actually requires, after you have produced three or four of them, you should be able to estimate quite accurately, and you will have hard quantitative evidence to use if you need to lobby for more time or more staff. Your boss may not know about writing manuals, but he or she does know about spreadsheets. You will be speaking a language that is understood in the business world — and that will make your request more credible. Implementing a scheduling system may not solve all your problems of short deadlines and multiple projects, but it will probably make meeting those deadlines a little less chaotic.

ORGANIZATIONAL SETTINGS THAT AFFECT WRITERS

The setting and organizational structure in which a writer operates can be the single most important factor in good manual production. Structures affecting manual writers vary enormously from company to company. These variations are sometimes attributable to the size of the company, to managerial philosophy, or to the maturity of the product. Here are some of the patterns that affect the manual writer's job.

Large Companies

The very large company typically has a divisional organization, a diversity of products at scattered geographic locations, and separate cadres of technical writers specializing in manuals for each product category. Further, by the time a company goes national or international, its product line is usually "mature"; i.e., the product has been around for some time, and the vocabulary for its parts and systems is quite well established and standardized. (Notable exceptions are the quick-growth electronic and computer technologies and those companies specializing in development of brand-new experimental products.)

Advantages

Very large companies with a team of technical writers whose sole responsibility is manual publication have the luxury of identifying and selecting good communicators from their own ranks. Alternatively, when they choose to hire new employees from outside, these companies usually have developed interviewing and testing systems to help them select the most-qualified applicants. Quite often, in large companies, writers come up through the ranks, transferring from parts or service manual writing or from positions in product safety, marketing, or advertising. They bring to the job an in-depth knowledge of the product. Large companies are also able, through their service publication managers and editors, to identify writers who need help with their writing skills. That help is provided by one-on-one editorial assistance, on-the-job orientation, and periodic training sessions.

The publication capabilities of large companies often exceed those found in the formal publishing world. Fully equipped photographic laboratories; sophisticated printing machines; computerized systems for layout, format, and translation and dedicated workstations; full-color duplicating machines; and in-house personnel specializing in art and technical drawing, slide production, film, and videotape — all of these are tools of the trade available at many large installations.

Disadvantages

There are also disadvantages to being a manual writer in a large company. Writers may have less autonomy and considerably less flexibility in deciding how best to do their job. If they are at widely scattered locations, they find that information takes longer to travel; filing systems may become harder to tap. If the large company is also decentralized, writing quality may be difficult to control. The manuals produced in Kentucky, for example, may be markedly different in quality and style from those produced in Florida. Because the large company tends to be more rigidly hierarchical, a decision to correct an error or to change the way manuals are done may take years, rather than months, to put into operation. In brief, what is gained through bigness, diversity, and sophistication may be lost through unwieldiness and lack of coordination.

Small and Intermediate-Size Companies

The small and intermediate-size company typically has one or only a few locations. Such companies tend to be regional and centralized and to have a limited product line. Quite often the product is young and innovative, and consequently there may be no old manuals to use as guides and no well-established vocabulary for parts and systems.

Advantages

For the writer, the small company can be an exciting and challenging place to work. A young product demands a fresh approach to the manual, and writers can literally create the vocabulary and the approach. Further, writers are less likely to have to deal with inertia or with "we've always done it this way" frustration. Designers and engineers are likely to be more accessible to answer questions, and decision making is usually more fluid and flexible because the small company hierarchy has fewer layers. In fact, some of the most inventive ideas for manual production and layout come from the small companies lucky enough to have creative writers who had to build a first-time manual from the ground up.

Disadvantages

Many small and intermediate-size companies assign the manual writing to a single individual or to a small group of writers. These writers may be confronted with an awesome array of tasks. They must learn the technology of the product, plan layout, write text and safety messages, arrange for art work, photographs, and drawings, negotiate with printers, edit, choose paper stock and typefaces — and often do their own typing and desktop publishing.

Publication support systems may be spotty in the small company (often little more than a typewriter, a desk top, and a corner in a office), and much of the production work must be contracted for. Manual writers who work "solo" feel the pressure of multiple responsibilities and are often rushed and isolated.

Understanding Internal Dynamics

Every company has its own internal peculiarities, its hierarchies, and pecking orders. Writers work within that pecking order, and situations will inevitably arise in which one person or unit has priority over another. Most writers can live comfortably with lines of authority, if they know what they are. What employees (writers included) find difficult are confused, pass-the-buck procedures in which the lines of authority are never articulated or clearly established and where hidden agendas dominate.

As a supervisor, you may know where final authority lies for decisions on the manual, but you may neglect to convey that information to your writers. You should try to let your writers know about situations in which their decisions are likely to be superseded by someone with higher authority.

In manual writing, the most common problems with lines of authority arise in the following procedures:

- Determining who will have final say and sign-off on the manual's technical accuracy
- Deciding on appropriate language levels for manuals (engineers and lawyers are often disturbed by the simplicity required for general public users)
- Deciding on final authority when distinctions must be made between legal safeguards and engineering safeguards
- Deciding on final authority when an editor and a writer disagree sharply on word choice, format, or stylistic preference

Whenever possible, decisions like these should be made by discussion and consensus, with writers included in the discussion. However, when negotiation is clearly not an option, let writers know where final authority lies.

When lines of authority are confusing or fuzzy, it's a good idea to follow up meetings with a memo to everyone in attendance outlining what you understood to be the decisions reached. That way, if you misinterpreted something, others can provide the correction. And if your interpretation is *not* disputed, you have written evidence that people were given the opportunity.

Writers must remember that technical publications sits in that odd position we discussed in Chapter 1: you are dependent on input from other departments, but no one (except the customer!) depends on your output. For that reason, it can get frustrating trying to get key people in other departments to act, either to provide information or sign off on manual text or illustrations. After all, your priorities are not their priorities. As we suggested in Chapter 1, use your networking skills to build relationships. You will be rewarded with cooperation and responsiveness.

SUMMARY

In Chapter 1, we discussed the writer's two basic needs: time and access to information. In this chapter, we added a third — training. If you are a manager of technical documentation or service publications, you should provide writers with as much assistance as possible. They should have

- Adequate time to do good work
- Access to information
- Training (in-house and/or outside)

We suggest that you reread the suggestions on information gathering and scheduling in Chapter 1. Many of the techniques suggested there involve tasks for which you, as a supervisor, may bear chief responsibility.

Most of the technical writers we talk with are eager to do competent work. They are also quick to sense whether management seems to be for or against them. As the supervisor of manual production, you should be your writers' chief advocate in insisting on training, information access, and time.

When a recent study of a government agency's documentation project asked writers and document designers what would help them do quality work, the suggestions included "better information about the audience, less rigid constraints, more control over the text, and less micro-management of the text by supervisors." The study also concluded that writers and designers "need adequate funding and a supportive atmosphere in which they can be creative." Without this support, companies and agencies suffer "missed opportunities as well as outright losses" in creating documents that reach the intended users (Schriver 202).

To the list of basic writers' needs we must now add a fourth: *recognition*. Times are changing. For many companies, the manual used to be regarded as a bothersome necessity that got written at the last minute by whoever was available to write it. Such attitudes were reflected in the scant time and money allotted to the manual and the meager recognition given to writers.

Now, a number of companies are realizing that the manual and the product should be designed together, in sync, for more effective documentation and product design. As companies move away from "over the wall" engineering and more toward some kind of concurrent engineering, they "need to embrace the idea that good communication starts with good planning. They need to move beyond the antiquated view of 'documentation as a nuisance activity' and bring their best communicators into the front end of product development" (Schriver 247).

Industries *are* beginning to recognize that manual writers are the bridge builders between the product and the consumer. After your product has left the dealer's store, the manual becomes the interpreter of your product. Without the manual, the consumer must make a phone call or a trip back to the seller for help.

As products grow more complex, as formerly simple mechanical devices are steadily being electronically controlled and computerized, and as manuals are interpreted by courts as significant legal evidence, the technical writer's work is becoming more valuable — and more highly valued. The new attitudes are reflected in better salaries, more investment in writing training programs, and better integration of the technical writer into the mainstream of company organizational structures.

The coalition of technical writers into a cohesive profession has also begun to take shape with academic programs and degrees, conferences, seminars, professional societies, newsletters, and books. The Society for Technical Communication, for example, is a national organization with regional and local chapters. Its publications and conferences help writers stay current in the field and keep in touch with each other.

We hope this book will prove to be another source of help, like the ones mentioned above and others listed elsewhere, to writers who have chosen technical writing as their profession.

Bibliography

ANSI Z535.4-1998, *American National Standard: Product Safety Signs and Labels,* National Electrical Manufacturers Association, Rosslyn, VA, 1998.

Baird, R. N., with McDonald, D. *The Graphics of Communication: Methods, Media, and Technology,* 6th edition, Harcourt Brace College Publishers, Ft. Worth, TX, 1993.

Bethune, James D. *Technical Illustration.* John Wiley & Sons, New York, 1983.

Bryan, Mark, with Julia Cameron and Catherine Allen. *The Artist's Way at Work*, William Morrow, New York, 1998.

Budiansky, Stephen, Lost in Translation, *The Atlantic*, December, 1998, 81–84.

Cameron, Julia. *The Artist's Way,* Jeremy P. Tarcher, New York, 1992.

Coe, Marlana, *Human Factors for Technical Communicators,* John Wiley & Sons, New York, 1996.

Coe, Marlana, Writing for Other Cultures: Ten Problem Areas, *Intercom*, vol. 44, no. 1, January, 1997, 17–19.

Cooper, Alan. The *Inmates Are Running the Asylum: Why High-Tech Products Drive Us Crazy and How to Restore the Sanity,* SAMS, a division of Macmillan Computer Publishing, Indianapolis, 1999.

Cornsweet, Tom. *Visual Perception.* Harcourt Brace College Publishers, Ft. Worth, TX, 1970.

Craig, J. and Meyer, S. E. (Eds.). *Designing with Type: A Basic Course in Typography,* Watson-Guptill, New York, 1999.

Dobb, F. P. *ISO 9000 Quality Registration*, Butterworth-Heinemann, Oxford, 1996, 1998.

Galosich, Allison. Operation HAACP, *The National Provisioner,* January, 1999.

Jastrzebski, Zbigniew. *Scientific Illustration.* Prentice-Hall, Englewood Cliffs, NJ, 1985.

Laux, Lila. A Human Factors Approach to Developing and Evaluating Facilitators. Presentations 1997–1999 at "The Role of Warnings and Instructions" seminars, University of Wisconsin–Madison.

Moll, Richard. Oral presentation at the University of Wisconsin Department of Engineering Professional Development, Madison, March 26, 1999.

Muir, J. and Gregg, T. *How to Keep Your Volkswagen Alive — A Manual of Step by Step Procedures for the Compleat Idiot*, John Muir Publications, Santa Fe, 1974.

Norman, Donald A. *The Design of Everyday Things,* Doubleday/Currency, New York, 1990. Originally published as *The Psychology of Everyday Things,* Basic Books, New York, 1988.

Perry, Susan K. *Writing in Flow,* Writer's Digest Books, Cincinnati, 1999.

Peters, G. A. and Peters, B. J. *Warnings, Instructions, and Technical Communications*, Lawyers and Judges Publishing Co., Tucson, AZ, 1999.

Ranous, C. A. Checklist developed at the University of Wisconsin–Madison.

Restatement of the Law Third, Torts: Products Liability. American Law Institute Publishers, St. Paul, MN, 1998.

Richardson, Graham T. *Illustrations*. Humana Press, Clifton, NJ, 1985.
Schriver, Karen A. *Dynamics in Document Design*, John Wiley & Sons, New York, 1997.
Siebert, L. and Ballard, L. *Making a Good Layout*, North Light Books, 1992.
Tufte, Edward R. *The Visual Display of Quantitative Information*, Graphics Press, Cheshire, CT, 1983.
Tufte, Edward R. *Visual Explanations*. Graphics Press, Cheshire, CT, 1997.
Wishbow, Nina. Home Sweet Home: Where Do Technical Communication Departments Belong? *Journal of Computer Documentation,* vol. 23, no. 1, February, 1999.

Index

A

B

C

D

I

J

L

M

N

O

P

Q

R

S

T

U

V

W